D0165574

TENSOR CALCULUS

A Concise Course

BARRY SPAIN
Sir John Cass College

DOVER PUBLICATIONS, INC.
Mineola, New York

Bibliographical Note

This Dover edition, first published in 2003, is an unabridged reprint of the third (1960) edition of *Tensor Calculus,* which was first published by Oliver and Boyd, Edinburgh and London, and by Interscience Publishers, Inc., New York, in 1953.

Library of Congress Cataloging-in-Publication Data

Spain, Barry.
Tensor calculus : a concise course / Barry Spain.
p. cm.
Originally published: 3rd ed., rev. Edinburgh : Oliver and Boyd ; New York: Interscience Publishers, 1960, in series: University mathematical texts.
Includes bibliographical references and index.
ISBN 0-486-42831-1 (pbk.)
1. Calculus of tensors. I. Title.

QA433.S65 2003
515'.63—dc21

2003041463

Manufactured in the United States of America
Dover Publications, Inc., 31 East 2nd Street, Mineola, N.Y. 11501

PREFACE

The object of this book is to provide a compact exposition of the fundamental results in the theory of tensors and also to illustrate the power of the tensor technique by applications to differential geometry, elasticity, and relativity. In the first five chapters the mathematical concepts are developed without undue stress on rigour. The remaining three chapters are independent of one another except that sections 38 and 39 of chapter VI, which treats Euclidean three-dimensional differential geometry, are necessary for a proper understanding of chapter VII which contains the theory of cartesian tensors and elasticity. Finally, chapter VIII is devoted both to the special and general theories of relativity. In the limited space available it is impossible to do justice to the physical principles underlying both these theories. But in order to help the reader unacquainted with relativity some explanatory matter has been incorporated into the text.

The presentation owes much to the authors listed in the bibliography, especially to McConnell, Synge and Schild. In particular, I wish to express my thanks to Dr. D. E. Rutherford for numerous suggestions and helpful criticisms during the manuscript and proof stages. Lastly I wish to thank my wife for her help with the proof-reading.

B. S.

Trinity College, Dublin; July, 1952

PREFACE TO THE THIRD EDITION

In this edition various errors have been corrected. Further, I wish to thank Mr. L. Lovitch for a neat proof (inserted on page 56) that the curvature tensor is zero in a flat space.

B. S.

CONTENTS

CHAPTER I

TENSOR ALGEBRA

CHAPTER II

THE LINE ELEMENT

CHAPTER III

COVARIANT DIFFERENTIATION

CHAPTER IV

GEODESICS - PARALLELISM

CHAPTER V

CURVATURE TENSOR

CHAPTER VI

EUCLIDEAN THREE-DIMENSIONAL DIFFERENTIAL GEOMETRY

CHAPTER VII

CARTESIAN TENSORS - ELASTICITY

CHAPTER VIII

THEORY OF RELATIVITY

TENSOR CALCULUS

A Concise Course

CHAPTER I

TENSOR ALGEBRA

§ 1. Introduction

The concept of a tensor has its origin in the developments of differential geometry by Gauss, Riemann and Christoffel. The emergence of Tensor Calculus, otherwise known as the Absolute Differential Calculus, as a systematic branch of Mathematics is due to Ricci and his pupil Levi-Civita. In collaboration they published the first memoir on this subject: — '*Methodes de calcul differential absolu et leurs applications*', *Mathematische Annalen*, vol. 54, (1901).

The investigation of relations which remain valid when we change from one coordinate system to any other, is the chief aim of Tensor Calculus. The laws of Physics can not depend on the frame of reference which the physicist chooses for the purpose of description. Accordingly it is aesthetically desirable and often convenient to utilise the Tensor Calculus as the mathematical background in which such laws can be formulated. In particular, Einstein found it an excellent tool for the presentation of his General Relativity theory. As a result the Tensor Calculus came into great prominence and is now invaluable in its applications to most branches of Theoretical Physics; it is also indispensable in the differential geometry of hyperspace.

It is assumed that the reader has an elementary knowledge of determinants and matrices. As he may not be acquainted with the Calculus of Variations, the minimum problem in the theory of geodesics is treated from first principles.

§ 2. N-Dimensional space

Consider an ordered set of N real variables $x^1, x^2, \ldots, x^i, \ldots, x^N$; these variables will be called the **coordinates** of a point. (The suffixes $1, 2, \ldots i, \ldots N$, which we shall call superscripts, merely serve as labels and do not possess any significance as power indices. Later we shall introduce quantities of the type a_i and again the i, which we shall call a subscript, will act only as a label.) Then all the points corresponding to all values of the coordinates are said to form an **N-dimensional space**, denoted by V_N. Several or all of the coordinates may be restricted in range to ensure a one-one correspondence between points of the V_N and sets of coordinates.

A **curve** in the V_N is defined as the assemblage of points which satisfy the N equations

$$x^i = x^i(u), \qquad (i = 1, 2, \ldots N)$$

where u is a parameter and $x^i(u)$ are N functions of u, which obey certain continuity conditions. In general, it will be sufficient that derivatives exist up to any order required.

A **subspace** V_M of V_N is defined for $M < N$ as the collection of points which satisfy the N equations

$$x^i = x^i(u^1, u^2, \ldots u^M), \qquad (i = 1, 2, \ldots N)$$

where there are M parameters $u^1, u^2, \ldots u^M$. The $x^i(u^1, u^2, \ldots u^M)$ are N functions of the $u^1, u^2, \ldots u^M$ satisfying certain conditions of continuity. In addition the $M \times N$ matrix formed from the partial derivatives $\partial x^i/\partial u^j$ is assumed to be of rank M*. When $M = N - 1$, the subspace is called a **hypersurface.**

§ 3. Transformation of coordinates

Let us consider a space V_N with the coordinate system $x^1, x^2, \ldots x^N$. The N equations

$$\bar{x}^i = \varphi^i(x^1, x^2, \ldots x^N), \qquad (i = 1, 2, \ldots N) \tag{3.1}$$

* T. Levi-Civita, The Absolute Differential Calculus, pp. 9—12.

where the φ^i are single-valued continuous differentiable functions of the coordinates, define a new coordinate system $\bar{x}^1, \bar{x}^2, \dots \bar{x}^N$. Equations (3.1) are said to define a **transformation of coordinates.** It is essential that the N functions φ^i be independent. A necessary and sufficient condition is that the Jacobian determinant formed from the partial derivatives $\partial\bar{x}^i/\partial x^j$ does not vanish*. Under this condition we can solve equations (3.1) for the x^i as functions of the $\bar{x}^i$ and obtain

$$x^i = \psi^i(\bar{x}^1, \bar{x}^2, \dots \bar{x}^N) \qquad (i = 1, 2, \dots N).$$

§ 4. Indicial and summation conventions

We will now introduce the following two conventions:
(1). *Latin indices, used either as subscripts or superscripts, will take all values from* 1 *to* N *unless the contrary is specified.* Thus equations (3.1) are briefly written $\bar{x}^i = \varphi^i(x^1, x^2, \dots x^N)$, the convention informing us that there are N equations.
(2). *If a Latin index is repeated in a term, then it is understood that a summation with respect to that index over the range* 1, 2, ... N *is implied.* Thus instead of the expression $\sum_{i=1}^{N} a_i x^i$, we merely write $a_i x^i$.

Now differentiation of (3.1) yields

$$d\bar{x}^i = \sum_{r=1}^{N} \frac{\partial\varphi^i}{\partial x^r}\, dx^r = \sum_{r=1}^{N} \frac{\partial\bar{x}^i}{\partial x^r}\, dx^r, \qquad (i = 1, 2, \dots N)$$

which simplify, when the above conventions are used, to

$$d\bar{x}^i = \frac{\partial\bar{x}^i}{\partial x^r}\, dx^r. \tag{4.1}$$

The repeated index r is called a **dummy index**, as it can be replaced by any other Latin index, except i in

* R. P. Gillespie, Partial Differentiation, pp. 43—46.

this particular case. That is, equations (4.1) can equally well be written $d\bar{x}^i = \dfrac{\partial \bar{x}^i}{\partial x^m}\, dx^m$ or for that matter $d\bar{x}^l = \dfrac{\partial \bar{x}^l}{\partial x^i}\, dx^i$. In order to avoid confusion, the same index must not be used more than twice in any single term. For example $(\sum_{i=1}^{N} a_i x^i)^2$ will not be written $a_i x^i a_i x^i$ but rather $a_i a_j x^i x^j$. It will always be clear from the context whether x^2 means x with superscript 2 or x squared. Usually powers will be indicated by the use of brackets; thus $(x^N)^2$ means the square of x^N. The reason for using superscripts and subscripts will be indicated in due course.

Let us introduce the **Kronecker delta** defined by

$$\begin{cases} \delta_j^k = 1 & \text{if } j = k \\ \delta_j^k = 0 & \text{if } j \neq k. \end{cases} \tag{4.2}$$

An obvious property of the Kronecker delta is that $\delta_j^k A^j = A^k$, since in the left-hand side of this equation the only surviving term is that for which $j = k$. Also $\partial x^k / \partial x^j = \delta_j^k$, because the coordinates x^i are independent.

Ex. Show that $\delta_j^i \delta_k^j = \delta_k^i$; $\delta_i^i = N$; $\dfrac{\partial x^k}{\partial \bar{x}^i} \dfrac{\partial \bar{x}^i}{\partial x^j} = \delta_j^k$.

§ 5. Contravariant vectors

A set of N functions A^i of the N coordinates x^i are said to be the components of a **contravariant vector** if they transform according to the equation.

$$\bar{A}^i = \frac{\partial \bar{x}^i}{\partial x^j} A^j, \tag{5.1}$$

on change of the coordinates x^i to $\bar{x}^i$. This means that any N functions can be chosen as the components of a contravariant vector in the coordinate system x^i, and the equations (5.1) define the N components in the new coordinate system $\bar{x}^i$. On multiplying equations (5.1) by

$\partial x^k/\partial \bar{x}^i$ and summing over the index i from 1 to N, we obtain

$$\frac{\partial x^k}{\partial \bar{x}^i} \bar{A}^i = \frac{\partial x^k}{\partial \bar{x}^i} \frac{\partial \bar{x}^i}{\partial x^j} A^j = \frac{\partial x^k}{\partial x^j} A^j = \delta^k_j A^j = A^k.$$

Hence the solution of equations (5.1) is

$$A^k = \frac{\partial x^k}{\partial \bar{x}^i} \bar{A}^i. \tag{5.2}$$

When we examine equations (4.1) we see that the differentials dx^i form the components of a contravariant vector, whose components in any other system are the differentials $d\bar{x}^i$ of that system. It follows immediately that dx^i/du is also a contravariant vector, called the **tangent vector** to the curve $x^i = x^i(u)$.

Consider now a further change of coordinates $x'^i = g^i(\bar{x}^1, \bar{x}^2, \ldots \bar{x}^N)$. Then the new components A'^i must be given by

$$A'^i = \frac{\partial x'^i}{\partial \bar{x}^j} \bar{A}^j = \frac{\partial x'^i}{\partial \bar{x}^j} \frac{\partial \bar{x}^j}{\partial x^k} A^k = \frac{\partial x'^i}{\partial x^k} A^k.$$

This equation is of the same form as (5.1), which shows that the transformations of contravariant vectors form a group *.

With the exception of the coordinates x^i themselves a single superscript will always denote a contravariant vector unless the contrary is explicitly stated. The coordinates x^i will only behave like the components of a contravariant vector with respect to linear transformations of the type $\bar{x}^i = a^i_j x^j$, where the a^i_j are a set of N^2 constants, which do not necessarily form the components of the entity introduced in section 8 and there called a tensor. For in this case $\partial \bar{x}^i/\partial x^j = a^i_j$ and the transformation can be rewritten $\bar{x}^i = \dfrac{\partial \bar{x}^i}{\partial x^j} x^j$. With respect to general transformations of coordinates, the x^i do not form

* W. Ledermann, Introduction to the Theory of Finite Groups, pp. 2—3.

the components of a contravariant vector. Essentially this means that if we select $A^i = x^i$, then the new components $\bar{A}^i$ with respect to the coordinate system $\bar{x}^i$ do *not* satisfy the equations $\bar{A}^i = \bar{x}^i$.

Ex. If a vector has components $\frac{d^2x}{dt^2}, \frac{d^2y}{dt^2}$ in rectangular cartesian coordinates, show that they are $\frac{d^2r}{dt^2} - r\left(\frac{d\theta}{dt}\right)^2, \frac{d^2\theta}{dt^2} + \frac{2}{r}\frac{dr}{dt}\frac{d\theta}{dt}$ in polar coordinates.

§ 6. Covariant vectors

A set of N functions A_i of the N coordinates x^i are said to be the components of a **covariant vector** if they transform according to the equation

$$\bar{A}_i = \frac{\partial x^j}{\partial \bar{x}^i} A_j, \tag{6.1}$$

on change of the coordinates x^i to $\bar{x}^i$. Any N functions can be chosen as the components of a covariant vector in the coordinate system x^i, and the equations (6.1) define the N components in the new coordinate system $\bar{x}^i$. On multiplying (6.1) by $\partial \bar{x}^i/\partial x^k$ and summing over the index i from 1 to N, we obtain

$$\frac{\partial \bar{x}^i}{\partial x^k} \bar{A}_i = \frac{\partial \bar{x}^i}{\partial x^k} \frac{\partial x^j}{\partial \bar{x}^i} A_j = \frac{\partial x^j}{\partial x^k} A_j = A_k. \tag{6.2}$$

Since $\frac{\partial f}{\partial \bar{x}^i} = \frac{\partial f}{\partial x^j} \frac{\partial x^j}{\partial \bar{x}^i}$, it follows immediately from (6.1) that the quantities $\partial f/\partial x^i$ are the components of a covariant vector, whose components in any other system are the corresponding partial derivatives $\partial f/\partial \bar{x}^i$. Such a covariant vector is called the **gradient** of f.

A single subscript will always denote a covariant vector unless the contrary is explicitly stated. In conformity with this convention we shall regard the index i in the covariant vector $\partial f/\partial x^i$ as a subscript.

We now show that there is no distinction between contravariant and covariant vectors when we restrict ourselves to transformations of the type

$$\bar{x}^i = a^i_m x^m + b^i, \tag{6.3}$$

where b^i are N constants which do not necessarily form the components of a contravariant vector and a^i_m are constants (not necessarily forming a tensor) such that

$$a^i_r a^i_m = \delta^r_m.$$

We multiply equations (6.3) by a^i_r and sum over the index i from 1 to N and obtain

$$x^r = a^i_r \bar{x}^i - a^i_r b^i.$$

Thus

$$\frac{\partial \bar{x}^i}{\partial x^j} = \frac{\partial x^j}{\partial \bar{x}^i} = a^i_j,$$

which shows that the equations (5.1) and (6.1) define the same type of entity.

Ex. Prove that the transformations of covariant vectors form a group.

§ 7. Invariants

Any function I of the N coordinates x^i is called an **invariant** or a **scalar** with respect to coordinate transformations if $I = \bar{I}$, where $\bar{I}$ is the value of I in the new coordinate system $\bar{x}^i$.

From the components A^i and B_i of a contravariant and covariant vector respectively, we can form the sum $A^i B_i$. When we change to new coordinates $\bar{x}^i$, this sum transforms to $\bar{A}^i \bar{B}_i$. Now

$$\bar{A}^i \bar{B}_i = \frac{\partial \bar{x}^i}{\partial x^j} A^j \frac{\partial x^k}{\partial \bar{x}^i} B_k = \delta^k_j A^j B_k = A^k B_k.$$

That is,

$$\bar{A}^i \bar{B}_i = A^i B_i.$$

Thus $A^i B_i$ is an invariant.

Another invariant is the quantity

$$\delta_i^i = \delta_1^1 + \delta_2^2 + \ldots + \delta_N^N = N.$$

§ 8. Second order tensors

Form the N^2 quantities $A^{ij} = B^i C^j$, where B^i and C^j are the components of two contravariant vectors. It follows from (5.1) that the A^{ij} transform according to

$$\bar{A}^{ij} = \frac{\partial \bar{x}^i}{\partial x^k} \frac{\partial \bar{x}^j}{\partial x^l} A^{kl}. \tag{8.1}$$

More generally, if we have N^2 functions A^{ij} whose transformation law is that of (8.1), then we call A^{ij} the components of a **contravariant tensor** of the **second order.** This tensor is not necessarily the product of the components of two contravariant vectors. Any set of N^2 functions can be chosen as the components of a contravariant tensor of the second order, and then (8.1) defines the components in any other coordinate system $\bar{x}^i$.

Similarly, if we have N^2 functions A_{ij} whose transformation law is

$$\bar{A}_{ij} = \frac{\partial x^k}{\partial \bar{x}^i} \frac{\partial x^l}{\partial \bar{x}^j} A_{kl}, \tag{8.2}$$

we call A_{ij} the components of a **covariant tensor** of the **second order.**

Further, if we have N^2 functions A_j^i whose transformation law is

$$\bar{A}_j^i = \frac{\partial \bar{x}^i}{\partial x^k} \frac{\partial x^l}{\partial \bar{x}^j} A_l^k, \tag{8.3}$$

we call A_j^i the components of a **mixed tensor** of the **second order.**

Note that the indices are placed on the tensors as superscripts when they denote contravariance and as subscripts when they denote covariance. In particular, the mixed tensor A_j^i transforms like a contravariant vector with respect to the index i and like a covariant vector with respect to the index j. Consequently the i is placed as a superscript whilst the j is placed as a subscript.

Now choose A^k_l to be the Kronecker delta δ^k_l. From (8.3) we have

$$\bar{A}^i_j = \frac{\partial \bar{x}^i}{\partial x^k}\frac{\partial x^l}{\partial \bar{x}^j}\delta^k_l = \frac{\partial \bar{x}^i}{\partial x^k}\frac{\partial x^k}{\partial \bar{x}^j} = \frac{\partial \bar{x}^i}{\partial \bar{x}^j} = \delta^i_j.$$

That is, the Kronecker delta is a mixed tensor of the second order whose components in any other system again form the Kronecker delta. This justifies the placing of one index as a subscript and the other as a superscript. Yet if we select the N^2 quantities $\delta_{ij} = \delta^i_j$ as the components of a covariant tensor in a coordinate system x^i, the components in the new system $\bar{x}^i$ are given by $\bar{\delta}_{ij} = \dfrac{\partial x^k}{\partial \bar{x}^i}\dfrac{\partial x^k}{\partial \bar{x}^j}$. Therefore the transformed components $\bar{\delta}_{ij}$ do not form the Kronecker delta.

Ex. Prove that $A_{ij}B^iC^j$ is an invariant, if B^i and C^j are contravariant vectors and A_{ij} a covariant tensor.

§ 9. Higher order tensors

A set of N^{s+p} functions $A^{t_1 t_2 \cdots t_s}_{q_1 q_2 \cdots q_p}$ of the N coordinates x^i are said to be the components of a **mixed tensor** of the $(s + p)$-th order, contravariant of the s-th order and covariant of the p-th order, if they transform according to the equation

$$(9.1)\quad \bar{A}^{u_1 u_2 \cdots u_s}_{r_1 r_2 \cdots r_p} = \frac{\partial \bar{x}^{u_1}}{\partial x^{t_1}} \cdots \frac{\partial \bar{x}^{u_s}}{\partial x^{t_s}}\frac{\partial x^{q_1}}{\partial \bar{x}^{r_1}} \cdots \frac{\partial x^{q_p}}{\partial \bar{x}^{r_p}} A^{t_1 t_2 \cdots t_s}_{q_1 q_2 \cdots q_p},$$

on change of the coordinates x^i to $\bar{x}^i$. This formula, although rather formidable in appearance, is merely a combination of (5.1) with respect to contravariant indices and of (6.1) with respect to covariant indices.

The order of indices in a tensor is important. The tensor A^{ij} is not necessarily the same as the tensor A^{ji}. (In the language of matrices A^{ji} is the transpose of A^{ij}.)

If however two contravariant indices or two covariant indices can be interchanged without altering the tensor, it is said to be **symmetric** with respect to these indices. We shall now prove that if a tensor is symmetric with respect to two indices in any coordinate system, it remains symmetric with respect to these two indices in any other coordinate system. There is no loss in generality in proving this for the contravariant tensor $A^{ij} = A^{ji}$. On applying (8.1) we have

$$(9.2) \qquad \bar{A}^{ij} = \frac{\partial \bar{x}^i}{\partial x^k} \frac{\partial \bar{x}^j}{\partial x^l} A^{kl} = \frac{\partial \bar{x}^j}{\partial x^l} \frac{\partial \bar{x}^i}{\partial x^k} A^{lk} = \bar{A}^{ji},$$

as required. We cannot usually define symmetry with respect to two indices, of which one denotes contravariance and the other covariance, because this symmetry may not be preserved after a coordinate transformation. The Kronecker delta, however, is a mixed tensor which possesses symmetry with respect to its two indices.

When *all* the indices of either a contravariant or a covariant tensor can be interchanged without altering the tensor, it is said to be symmetric. A symmetric tensor of the second order has at most $\frac{1}{2}N(N + 1)$ different components.

A tensor each component of which alters in sign but not in magnitude when two contravariant indices or two covariant indices are interchanged is said to be **skew-symmetric** with respect to these indices. It can be shown by equations similar to (9.2) that the property of skew-symmetry is also independent of the choice of the coordinate system. Skew-symmetry, like symmetry, cannot be defined with respect to two indices, of which one denotes contravariance and the other covariance.

If *all* the indices of a contravariant or a covariant tensor can be interchanged so that the tensor changes its sign at each interchange of a pair of indices, the tensor is said to be skew-symmetric. A skew-symmetric tensor A^{ij} of the second order has at most $\frac{1}{2}N(N - 1)$ different *arithmetical* components, as all the quantities A^{ii} (no summation) are zero. The several components of a skew-symmetric tensor of the N-th order are either zero or differ merely in sign. So there is essentially only one non-zero component of such a tensor.

The most important deduction from (9.1) is this: if all the components of a tensor in one coordinate system are zero at a point, they are all zero at this point in every coordinate system. Further, if the components are identically zero in one coordinate system, they are also identically zero in every coordinate system. It is this property which constitutes the importance of tensors in physical applications.

When a tensor is defined at all points of a curve or throughout the space V_N itself, we say that it constitutes a tensor-field.

Ex. 1. If A_{ij} is a skew-symmetric tensor, prove that

$$(\delta^i_j \delta^k_l + \delta^i_l \delta^k_j) A_{ik} = 0.$$

Ex. 2. Prove that the transformations of tensors form a group.

§ 10. Addition, subtraction and multiplication of tensors

It is clear that we cannot expect to give any tensorial meaning to the expression $A^{ij} + B^i$, because it cannot satisfy the transformation law (9.1). It does, however, follow from this equation that any linear combination of tensors of the *same* type whose coefficients are invariants is a tensor of the *same* type. For example, from the two tensors A^i_{jk} and B^i_{jk}, we can form the tensor $\lambda A^i_{jk} + \mu B^i_{jk}$ which will satisfy (9.1) provided that λ and μ are invariants. In particular, $A^i_{jk} + B^i_{jk}$ and $A^i_{jk} - B^i_{jk}$ are called the **sum** and **difference** respectively of the two tensors. As another example, we can write

$$A_{ij} = \tfrac{1}{2}(A_{ij} + A_{ji}) + \tfrac{1}{2}(A_{ij} - A_{ji}).$$

Now $A_{ij} + A_{ji}$ is symmetric and $A_{ij} - A_{ji}$ is skew-symmetric. Thus any covariant tensor of the second order is the sum of a symmetric and a skew-symmetric tensor. This is, of course, also true of a contravariant tensor of the second order.

Let us select two tensors, one of contravariant order s and covariant order p, the other of contravariant order t and covariant order q. It then follows from (9.1) that the products of the components form a mixed tensor of contravariant order $s + t$ and covariant order $p + q$. This tensor is called the **outer product** of the two tensors. For instance $A^{ijl}_{kmnt} = B^{ij}_{k} C^{l}_{mnt}$ is the outer product of the two tensors B^{ij}_{k} and C^{l}_{mnt} and is a tensor of the type indicated by its indices.

The division, in the usual sense, of one tensor by another is not defined.

§ 11. Contraction

Let us start with any mixed tensor, say A^{ij}_{lmn}, and form the sum A^{ij}_{lmj}. From (9.1) we have

$$\bar{A}^{st}_{pqr} = \frac{\partial \bar{x}^s}{\partial x^i} \frac{\partial \bar{x}^t}{\partial x^j} \frac{\partial x^l}{\partial \bar{x}^p} \frac{\partial x^m}{\partial \bar{x}^q} \frac{\partial x^n}{\partial \bar{x}^r} A^{ij}_{lmn}.$$

Therefore

$$\bar{A}^{sr}_{pqr} = \frac{\partial \bar{x}^s}{\partial x^i} \frac{\partial \bar{x}^r}{\partial x^j} \frac{\partial x^l}{\partial \bar{x}^p} \frac{\partial x^m}{\partial \bar{x}^q} \frac{\partial x^n}{\partial \bar{x}^r} A^{ij}_{lmn}$$

$$= \frac{\partial \bar{x}^s}{\partial x^i} \frac{\partial x^l}{\partial \bar{x}^p} \frac{\partial x^m}{\partial \bar{x}^q} \delta^n_j A^{ij}_{lmn}$$

$$= \frac{\partial \bar{x}^s}{\partial x^i} \frac{\partial x^l}{\partial \bar{x}^p} \frac{\partial x^m}{\partial \bar{x}^q} A^{in}_{lmn}.$$

Thus we observe that A^{ij}_{lmj} is a mixed tensor, contravariant of the first order and covariant of the second order. This process, which is called **contraction**, enables us to obtain a tensor of order $r - 2$ from a mixed tensor of order r. In the above example we could contract a stage further and arrive at the covariant vector A^{ij}_{lij}. When contracting, any superscript may be used to sum with any subscript. Therefore we can form the following different tensors by contraction:- A^{ij}_{lmj}, A^{ij}_{ljn}, A^{ij}_{jmn}, A^{ij}_{lmi}, A^{ij}_{lin}, A^{ij}_{imn}, A^{ij}_{lij}, A^{ij}_{lji}, A^{ij}_{imj}, A^{ij}_{jmi}, A^{ij}_{ijn} and A^{ij}_{jin}. If the tensor A^{ij}_{lmn} possesses any symmetric properties, there will be fewer tensors formed from it by contraction. As another example, the invariant A^i_i is formed by contraction from the mixed tensor A^i_j. This justifies us in calling an invariant a tensor of zero order.

We can also combine multiplication and contraction to produce new tensors. From the tensors A^{ij}_k and B^l_{mnt}, we may obtain such tensors as $A^{ij}_k B^k_{mnt}$, $A^{ij}_k B^l_{inj}$, $A^{ij}_k B^k_{mji}$ and many others. This process is called **inner multiplication** of two tensors and the resulting tensor is called an **inner product** of the two tensors.

Note carefully that we never contract two indices of the same type as the resulting sum is not necessarily a tensor. Also it should now be clear that with our index notation, the summation convention generally applies to two indices one of which is a superscript and the other a subscript.

§ 12. Quotient law

Sometimes it is necessary to ascertain whether a set of functions form the components of a tensor. The direct method requires us to find out if they satisfy a tensor transformation equation of the type (9.1). In practice this is troublesome and a simpler test is provided by the quotient law. The **quotient law** states that N^p functions of x^i form the components of a tensor of order p, (whose contravariant and covariant character can readily be determined), provided that an inner product of these functions with an *arbitrary* tensor is itself a tensor. It will suffice to set out the proof for the following particular case. The set of N^3 functions A^{ijk} form the components of a tensor of the type indicated by its indices if

$$A^{ijk} B^p_{ij} = C^{pk},$$

provided that B^p_{ij} is an *arbitrary* tensor and C^{pk} a tensor. The transformed quantities, referred to a system of coordinates $\bar{x}^i$, satisfy the equations

$$\bar{A}^{ijk} \bar{B}^p_{ij} = \bar{C}^{pk},$$

which become on substitution from (9.1)

$$\bar{A}^{ijk} \frac{\partial \bar{x}^p}{\partial x^l} \frac{\partial x^m}{\partial \bar{x}^i} \frac{\partial x^n}{\partial \bar{x}^j} B^l_{mn} = \frac{\partial \bar{x}^p}{\partial x^q} \frac{\partial \bar{x}^k}{\partial x^r} C^{qr}$$

$$= \frac{\partial \bar{x}^p}{\partial x^q} \frac{\partial \bar{x}^k}{\partial x^r} A^{ijr} B^q_{ij}.$$

With a change of dummy indices, we have

$$\frac{\partial \bar{x}^p}{\partial x^l} \left[\bar{A}^{ijk} \frac{\partial x^m}{\partial \bar{x}^i} \frac{\partial x^n}{\partial \bar{x}^j} - A^{mnr} \frac{\partial \bar{x}^k}{\partial x^r} \right] B^l_{mn} = 0.$$

On multiplying this equation by $\partial x^s / \partial \bar{x}^p$ and summing over p from 1 to N, (in future we shall merely write 'on inner multiplication by $\partial x^s / \partial \bar{x}^p$') we obtain

$$\left[\bar{A}^{ijk} \frac{\partial x^m}{\partial \bar{x}^i} \frac{\partial x^n}{\partial \bar{x}^j} - A^{mnr} \frac{\partial \bar{x}^k}{\partial x^r} \right] B^s_{mn} = 0. \tag{12.1}$$

Since B^s_{mn} is an arbitrary tensor, we can arrange that only one of its components differs from zero. Now each component of B^s_{mn} may be selected in turn as that one which does not vanish. This shows that the expression in brackets is identically zero. That is

$$\bar{A}^{ijk} \frac{\partial x^m}{\partial \bar{x}^i} \frac{\partial x^n}{\partial \bar{x}^j} = A^{mnr} \frac{\partial \bar{x}^k}{\partial x^r}. \tag{12.2}$$

Inner multiplication of this equation by $\dfrac{\partial \bar{x}^s}{\partial x^m} \dfrac{\partial \bar{x}^t}{\partial x^n}$ yields the result

$$\bar{A}^{stk} = \frac{\partial \bar{x}^s}{\partial x^m} \frac{\partial \bar{x}^t}{\partial x^n} \frac{\partial \bar{x}^k}{\partial x^r} A^{mnr}.$$

Thus A^{mnr} is a tensor of the third order and contravariant in all its indices. In the above proof it is important that the tensor B^l_{mn} shall be arbitrary and must not possess any symmetric or skew-symmetric properties. Let us examine what happens if B^l_{mn} is symmetrical in m and n. It is no longer possible to deduce (12.2) from (12.1). The correct deduction now is that

$$\bar{A}^{ijk}\frac{\partial x^m}{\partial \bar{x}^i}\frac{\partial x^n}{\partial \bar{x}^j} + \bar{A}^{ijk}\frac{\partial x^n}{\partial \bar{x}^i}\frac{\partial x^m}{\partial \bar{x}^j} = A^{mnr}\frac{\partial \bar{x}^k}{\partial x^r} + A^{nmr}\frac{\partial \bar{x}^k}{\partial x^r}.$$

On changing several dummy indices, this equation becomes

$$(\bar{A}^{ijk} + \bar{A}^{jik})\frac{\partial x^m}{\partial \bar{x}^i}\frac{\partial x^n}{\partial \bar{x}^j} = (A^{mnr} + A^{nmr})\frac{\partial \bar{x}^k}{\partial x^r}.$$

Inner multiplication by $\dfrac{\partial \bar{x}^s}{\partial x^m}\dfrac{\partial \bar{x}^t}{\partial x^n}$ shows that $A^{mnr} + A^{nmr}$ is a contravariant tensor of the third order. If in addition we know that A^{mnr} is symmetric in m and n, it follows immediately that A^{mnr} is a symmetric tensor with respect to the indices m and n. We observe, therefore, that the quotient law must be applied with care.

Ex. 1. If A^i and B^i are arbitrary contravariant vectors and $C_{ij}A^iB^j$ is an invariant, show that C_{ij} is a covariant tensor of the second order.

Ex. 2. If A^i is an arbitrary contravariant vector and $C_{ij}A^iA^j$ is an invariant, show that $C_{ij} + C_{ji}$ is a covariant tensor of the second order.

§ 13. Conjugate symmetric tensors of the second order

Consider a symmetric covariant tensor of the second order A_{ij} whose determinant $|A_{ij}| \neq 0$. Let B^{ij} denote the expression formed by dividing the cofactor of A_{ij} in the determinant $|A_{ij}|$ by $|A_{ij}|$ itself. We shall prove that the B^{ij} so obtained are the components of a contravariant tensor of the second order. In anticipation of this, B^{ij} is labelled as if it were a contravariant tensor. We have from the theory of determinants*.

$$A_{ij} B^{ik} = \delta_j^k. \tag{13.1}$$

We cannot establish the tensor character of B^{ik} by applying the quotient law directly to this equation, because A_{ij} is not arbitrary. Let us choose an arbitrary contravariant vector C^i. Then $D_i = A_{ij}C^j$ is an arbitrary covariant vector, because these N equations can be uniquely solved for the C^i in terms of the D_i since $| A_{ij} | \neq 0$. Consequently,

$$D_i B^{ik} = A_{ij} C^j B^{ik} = \delta_j^k C^j = C^k.$$

Now if we apply the quotient law to the equation $D_i B^{ik} = C^k$, we see that B^{ik} is a contravariant tensor of the second order. Also it is clear from the definition that B^{ij}, like A_{ij}, is symmetric.

We will next attempt by the same process to obtain another tensor from B^{ik}. Let E_{ij} denote the cofactor of B^{ij} in the determinant $| B_{ij} |$ divided by $| B_{ij} |$ itself. From the theory of determinants $| A_{ij} | . | B_{ij} | = 1$ and consequently $| B_{ij} | \neq 0$, which means that E_{ij} always exists. Further we have

$$E_{ij} B^{ik} = \delta_j^k.$$

Inner multiplication by A_{kl} yields on application of (13.1) that

$$E_{lj} = A_{jl} = A_{lj}.$$

Thus this process only leads back to the original covariant tensor of the second order. We say that A_{ij} and B^{ij} are **conjugate** tensors if they satisfy equations (13.1). It is important to note that a tensor of the second order has a conjugate only if its determinant is not zero.

Ex. If $A_{ij} = 0$ for $i \neq j$, show that the conjugate tensor $B^{ij} = 0$ for $i \neq j$, and $B^{ii} = 1/A_{ii}$ (no summation).

* A. C. Aitken, Determinants and Matrices, pp. 51—52.

CHAPTER II

THE LINE ELEMENT

§ 14. Fundamental tensor

At this stage we introduce the concept of distance into our space V_N. If the distance ds between the neighbouring points with coordinates x^i and $x^i + dx^i$ is given by the quadratic differential form

$$ds^2 = g_{ij}\,dx^i\,dx^j \tag{14.1}$$

where the g_{ij} are functions of x^i, subject only to the restriction $g = | g_{ij} | \neq 0$, the space is said to be a **Riemannian space.** In addition we postulate that the distance between two neighbouring points is independent of the coordinate system. That is, ds is an invariant. From the quotient law, it follows that $g_{ij} + g_{ji}$ is a covariant tensor of the second order. We can write

$$g_{ij} = \tfrac{1}{2}(g_{ij} + g_{ji}) + \tfrac{1}{2}(g_{ij} - g_{ji}).$$

The contribution of $\frac{1}{2}(g_{ij} - g_{ji})dx^i dx^j$ to ds^2 is zero, hence there is no loss of generality in assuming that g_{ij} is symmetric. Thus g_{ij} is a covariant symmetrical tensor of the second order called the **fundamental tensor** of the Riemannian space. The quadratic form $g_{ij}dx^i dx^j$ is called the **metric.** It is also the square of the **line-element** ds.

The line-element ds of a three dimensional Euclidean space, referred to a system of rectangular cartesian axes, is

$$ds^2 = (dx^1)^2 + (dx^2)^2 + (dx^3)^2.$$

All the components of the fundamental tensor are zero except $g_{11} = g_{22} = g_{33} = 1$. It is evident that the metric of a Euclidean space is positive-definite*. That is, ds^2 is zero when $dx^1 = dx^2 = dx^3 = 0$ but can only take positive values for all other *real* values of dx^1, dx^2 and dx^3.

The Special Theory of Relativity discusses the four dimensional space with the line-element ds given by

$$(14.2) \quad ds^2 = -(dx^1)^2 - (dx^2)^2 - (dx^3)^2 + c^2(dx^4)^2$$

This metric is not positive-definite, as it is positive for all curves along which x^1, x^2 and x^3 are all constants, but negative for all curves along which x^4 is constant. Thus along these latter curves the distance between neighbouring points cannot be real. In order that the distance ds between two neighbouring points be real, equation (14.1) will be amended to

$$(14.3) \qquad ds^2 = eg_{ij}\,dx^i\,dx^j,$$

where the factor e called the **indicator** takes the value $+1$ or -1 so that ds^2 is always positive.

Ex. Show that the metric of a Euclidean space, referred to spherical polar coordinates $x^1 = r$, $x^2 = \theta$ and $x^3 = \psi$ is given by $ds^2 = dr^2 + r^2 d\theta^2 + r^2 \sin^2 \theta d\psi^2$.

§ 15. Length of a curve

Consider the curve $x^i = x^i(t)$ with the parameter t. From (14.3) the length of the curve between the points corresponding to $t = t_1$ and $t = t_2$ is given by

$$(15.1) \qquad s = \int_{t_1}^{t_2} \sqrt{eg_{ij} \frac{dx^i}{dt} \frac{dx^j}{dt}}\, dt.$$

If $g_{ij} \dfrac{dx^i}{dt} \dfrac{dx^j}{dt} = 0$ along a curve, then the two points corresponding to t_1 and t_2 are at zero distance from one another, despite the fact that they are not coincident.

* A. J. Mc. Connell, Absolute Differential Calculus, p. 16.
A. C. Aitken, Determinants and Matrices, p. 137.

Such a curve is called **minimal** or **null.** The curve given by

$$(15.2)\quad \begin{cases} x^1 = c\int r\cos\theta\cos\psi\, dt, & x^2 = c\int r\cos\theta\sin\psi\, dt, \\ x^3 = c\int r\sin\theta\, dt, & x^4 = \int r dt, \end{cases}$$

where r, θ and ψ are functions of t is a *real* null curve in the V_4 whose metric is (14.2). It is clear that no real null curves lie in a Riemannian space whose metric is positive-definite.

A curve will consist of portions along which the indicator e is $+1$, portions along which the indicator is -1, and null portions. The length of the curve is then the sum of the lengths of these portions, the null part contributing zero to the value of the length.

Except in the case of null curves, the parameter t may be chosen as the arc-distance s from some fixed point of the curve. From (14.3)

$$(15.3)\qquad g_{ij}\frac{dx^i}{ds}\frac{dx^j}{ds} = e$$

along any portion of a curve which is not null.

§ 16. Magnitude of a vector

The **magnitude** A of the contravariant vector A^i is defined by

$$(16.1)\qquad (A)^2 = e_{(A)}\, g_{ij}A^iA^j,$$

where $e_{(A)}$ is the indicator $+1$ or -1 which makes A real. The magnitude A is an invariant. In a Euclidean space, referred to rectangular cartesian coordinates, (16.1) reduces to the familiar definition of the magnitude of a vector.

At this stage it is necessary to introduce the contravariant tensor conjugate to g_{ij}, which can conveniently be written g^{ij}. Then equations (13.1) read

$$(16.2)\qquad g_{ij}g^{ik} = \delta_j^k.$$

We can now define the magnitude B of the covariant vector B_i by the equation

$$(16.3) \qquad (B)^2 = e_{(B)}\, g^{ij} B_i B_j,$$

where $e_{(B)}$ is the indicator of the vector B_i, and it is clear that B is an invariant.

A vector, whose magnitude is unity, is called a **unit vector.** It follows from (15.3) that dx^i/ds is a unit contravariant vector. If the magnitude of a vector is zero, it is called a **null vector.** The tangent vector to a null curve is a null vector.

§ 17. Associate tensors

The inner product of the fundamental tensor g_{ij} and the contravariant vector A^j is the covariant vector $g_{ij}A^j$, which is said to be **associate** to A^j. We define

$$(17.1) \qquad A_i = g_{ij}A^j.$$

Similarly we define

$$B^i = g^{ij} B_j$$

and say that the vector B^i is associate to the vector B_j.

The relation between a vector and its associate is reciprocal, for the vector associate to A_i is

$$g^{ij}A_j = g^{ij} g_{jk} A^k = \delta^i_k A^k = A^i.$$

This process of association is often referred to as 'lowering the superscript' or 'raising the subscript' respectively. We have

$$e_{(A)}\, g_{ij} A^i A^j = e_{(A)} g_{ij} g^{ik} A_k\, g^{jl} A_l = e_{(A)} g^{kl} A_k A_l,$$

which shows that the magnitudes of associate vectors are equal.

The process of raising and lowering indices can be performed on tensors. From the tensor $A^{ijk}{}_{\dots lm}$ we can form associate tensors like $A_{r\cdot\cdot lm}^{\cdot jk} = g_{ri} A^{ijk}{}_{\dots lm}$ or $A^{i\cdot kst}_{\cdot r} = g_{jr} g^{ls} g^{mt} A^{ijk}{}_{\dots lm}$.

The dot notation is introduced to indicate which indices have been raised or lowered. The dots will be omitted when there is no possibility of confusion. For example, we shall write $A^{ij} = g^{ir}g^{js}A_{rs}$. Note very carefully that although g_{ij} and g^{ij} are conjugate tensors, the tensors A_{ij} and A^{ij} are not as a rule conjugate.

Ex. Show that $(A)^2 = e_{(A)}A_iA^i$.

§ 18. Angle between two vectors - orthogonality

The angle between two *unit* vectors A^i and B^i is defined by

$$\text{(18.1)} \quad \cos\theta = g_{ij}A^iB^j = A_jB^j = g^{jk}A_jB_k = A^kB_k.$$

This equation will be found to agree with the usual formula $\cos\theta = ll' + mn' + nn'$ for the angle between the unit vectors (l, m, n) and (l', m', n') in a three dimensional Euclidean space when it is referred to rectangular cartesian coordinates. We will now show that (18.1) always defines a real angle between two real vectors if the metric of the Riemannian space is *positive-definite*. For then the magnitude of the vector $\lambda A^i + \mu B^i$ is greater than or equal to zero for all real values of λ and μ. That is

$$g_{ij}(\lambda A^i + \mu B^i)(\lambda A^j + \mu B^j) \geqq 0,$$

which reduces to

$$\lambda^2 + 2\lambda\mu\cos\theta + \mu^2 \geqq 0.$$

That is

$$(\lambda + \mu\cos\theta)^2 + \mu^2(1 - \cos^2\theta) \geqq 0.$$

Since this equation is true for all values of λ and μ, it follows that $1 - \cos^2\theta \geqq 0$. Thus $|\cos\theta| \leqq 1$ which means that θ is real. If the metric is not positive-definite then the angle between two real unit vectors need not be real.

An immediate deduction from (18.1) is that the angle θ between two vectors A^i and B^i, which are not necessarily unit vectors, is given by

$$\cos\theta = \frac{g_{ij}A^i B^j}{\sqrt{e_{(A)}e_{(B)}g_{lm}A^l A^m g_{rs}B^r B^s}}. \tag{18.2}$$

Two vectors are said to be **orthogonal** to one another if the angle between them is a right angle. From (18.2) the necessary and sufficient condition for the orthogonality of the two vectors A^i and B^i is that

$$g_{ij}A^i B^j = 0. \tag{18.3}$$

We do not define the angle between two vectors if one or both of them happens to be a null vector, but we shall take (18.3) as the definition of orthogonality of two null vectors. It follows that a null vector is self-orthogonal.

Ex. Prove that (1, 0, 0, 0) and ($\sqrt{2}$, 0, 0, $\sqrt{3}/c$) are unit vectors in the V_4 with the metric (14.2). Show also that the angle between these vectors is not real.

§ 19. Principal directions

From the symmetric covariant tensor A_{ij} we can construct the determinantal equation

$$|A_{ij} - \lambda g_{ij}| = 0, \tag{19.1}$$

which is of degree N in λ. When we change to a new coordinate system $\bar{x}^i$, this equation transforms to

$$\left| (\bar{A}_{lk} - \lambda \bar{g}_{lk}) \frac{\partial \bar{x}^l}{\partial x^i} \frac{\partial \bar{x}^k}{\partial x^j} \right| = 0.$$

On applying the multiplication law of determinants, this equation can be written

$$\left| \bar{A}_{lk} - \lambda \bar{g}_{lk} \right| \cdot \left| \frac{\partial \bar{x}^l}{\partial x^i} \right|^2 = 0.$$

The Jacobian determinant $\left| \frac{\partial \bar{x}^l}{\partial x^i} \right|$ does not vanish; consequently this equation reduces to

$$|\bar{A}_{lk} - \lambda \bar{g}_{lk}| = 0.$$

By comparison with (19.1), we deduce that the roots $\lambda_{(K)}$ of this equation are invariants. (Capital letters are used to label the roots. Their enclosure in brackets emphasises that they have no tensorial significance, and the summation convention is not to apply to them.)

Now consider the N equations

$$(A_{ij} - \lambda_{(K)} g_{ij}) L^i_{(K)} = 0, \tag{19.2}$$

where $\lambda_{(K)}$ is a simple root of (19.1). These determine the ratios of the N values of $L^i_{(K)}$. We cannot deduce immediately from the quotient law that $L^i_{(K)}$ is a contravariant vector, because the tensor $A_{ij} - \lambda_{(K)} g_{ij}$, with which it shares inner multiplication, is not arbitrary. Instead we change to the coordinate system $\bar{x}^i$, and equations (19.2) transform into

$$(\bar{A}_{lk} - \lambda_{(K)} \bar{g}_{lk}) \frac{\partial \bar{x}^l}{\partial x^i} \frac{\partial \bar{x}^k}{\partial x^j} L^i_{(K)} = 0.$$

Inner multiplication by $\partial x^j / \partial \bar{x}^m$ yields

$$(\bar{A}_{lm} - \lambda_{(K)} \bar{g}_{lm}) \frac{\partial \bar{x}^l}{\partial x^i} L^i_{(K)} = 0.$$

These N equations determine the ratios of the N quantities $\dfrac{\partial \bar{x}^l}{\partial x^i} L^i_{(K)}$, which are the components of $L^l_{(K)}$ in the $\bar{x}^i$ coordinate system. That is, $L^i_{(K)}$ transforms in accordance with equations (5.1) and thus $L^i_{(K)}$ is a contravariant vector. Let us now choose the ratios of the $L^i_{(K)}$ so that it is a unit vector. That is,

$$g_{ij} L^i_{(K)} L^j_{(K)} = e_{(K)}, \tag{19.3}$$

where $e_{(K)}$ is the indicator of the vector $L^i_{(K)}$.

Similarly we can show that corresponding to each *simple* root $\lambda_{(M)}$ of equation (19.1) there corresponds the unit vector $L^i_{(M)}$ satisfying the equations

$$(A_{ij} - \lambda_{(M)} g_{ij}) L^i_{(M)} = 0 \tag{19.4}$$

and

$$g_{ij}L^i_{(M)}L^j_{(M)}) = e_{(M)}. \tag{19.5}$$

When $\lambda_{(M)}$ is a *multiple* root of (19.1), the vector $L^i_{(M)}$ is not uniquely determined by equations (19.4) and (19.5).

Let us choose two simple roots $\lambda_{(K)}$ and $\lambda_{(M)}$ of (19.1). Since $\lambda_{(K)} \neq \lambda_{(M)}$, the inner multiplication of (19.2) by $L^j_{(M)}$ and (19.4) by $L^j_{(K)}$ gives us on subtraction, the equation

$$g_{ij}L^i_{(K)}L^j_{(M)} = 0. \tag{19.6}$$

This shows that the two unit vectors $L^i_{(K)}$ and $L^i_{(M)}$ are orthogonal. Thus if all the roots of equation (19.1) are simple at any point, the covariant symmetric tensor determines uniquely N mutually orthogonal unit vectors. The directions of these vectors at a point are called the **principal directions** determined by A_{ij}. If the metric $g_{ij}\,dx^i\,dx^j$ is positive-definite all the roots of (19.1) are real*. That is, the principal directions are real in a space with positive-definite metric.

Now consider the invariant λ defined by

$$\lambda = \frac{A_{ij}L^iL^j}{g_{lm}L^lL^m}.$$

The finite maxima and minima of λ are given by $\dfrac{\partial\lambda}{\partial L^j} = 0$, that is, by

$$A_{ij}L^i(g_{lm}L^lL^m) - g_{ij}L^i(A_{lm}L^lL^m) = 0.$$

This equation can be written

$$(A_{ij} - \lambda g_{ij})L^i = 0.$$

Elimination of the L^i yields the determinantal equation

$$| A_{ij} - \lambda g_{ij} | = 0.$$

Thus the finite maxima and minima of λ are those values corresponding to the principal directions determined by A_{ij}.

* W. L. Ferrar, Algebra, p. 145.

If $A_{ij} = \lambda g_{ij}$ at a point, then the principal directions are indeterminate at that point. If $A_{ij} \equiv \lambda g_{ij}$ at all points of a V_N, the space is said to be **homogeneous** with respect to the tensor A_{ij}.

In a Euclidean space of N dimensions referred to rectangular cartesian coordinates, the components of the fundamental tensor g_{ij} form the unit matrix. Hence the roots of (19.1) are in this case the latent roots* of the matrix A_{ij} and the principal directions are those of the latent vectors.

Ex. Prove that $A_{ij} L^i_{(K)} L^j_{(K)} = e_{(K)} \lambda_{(K)}$ and $A_{ij} L^i_{(K)} L^j_{(M)} = 0$. (No summation over (K)).

Examples

1. Show that the angle θ between the vectors A^i and B^i is given by

$$\sin^2 \theta = \frac{(e_{(A)} e_{(B)} g_{hi} g_{jk} - g_{hk} g_{ij}) A^h A^i B^j B^k}{e_{(A)} e_{(B)} g_{hi} g_{jk} A^h A^i B^j B^k}.$$

2. If A_{ij} is a skew-symmetric covariant tensor, prove that $(A_{23}/\sqrt{g}, A_{31}/\sqrt{g}, A_{12}/\sqrt{g})$ are the components of a contravariant vector. Show also that $(\sqrt{g}A^{23}, \sqrt{g}A^{31}, \sqrt{g}A^{12})$ are the components of a covariant vector if A^{ij} is a skew-symmetric contravariant tensor.

3. Prove that no relation of the type

$$\lambda g_{ij} A_{hk} + \mu g_{ih} A_{jk} + \nu g_{ik} A_{hj} = 0$$

can exist in a V_N, $(N > 1)$, where λ, μ and ν are invariants and A_{hk} is a symmetric tensor.
If A_{hk} is a skew-symmetric tensor, show that this equation can not exist in a V_N for which $N > 2$.

* A. C. Aitken, Determinants and matrices, p. 73.

CHAPTER III

COVARIANT DIFFERENTIATION

§ 20. Christoffel symbols

Although we found in section 5 that dx^i/du is always a contravariant vector, the exercise at the end of that section should convince us that its derivatives d^2x^i/du^2 do *not* form a vector, whose components in any other system are the corresponding second derivatives. Again, in section 6 we proved that the partial derivatives of an invariant form the components of a covariant vector; yet we shall show in section 22 that the derivatives of a vector do not form a tensor whose components in any other system are the corresponding derivatives of the transformed vector. Our aim is now to build up expressions involving the derivatives of a tensor, which are the components of a tensor. In order to carry out this programme we must first investigate two functions formed from the fundamental tensor g_{ij}. These are the Christoffel symbols of the first and second kinds defined respectively by

$$(20.1) \qquad [ij,\, k] = \frac{1}{2}\left(\frac{\partial g_{ik}}{\partial x^j} + \frac{\partial g_{jk}}{\partial x^i} - \frac{\partial g_{ij}}{\partial x^k}\right)$$

and

$$(20.2) \qquad \begin{Bmatrix} l \\ ij \end{Bmatrix} = g^{lk}[ij,\, k].$$

Although we shall see that the symbols $[ij,\, k]$ and $\begin{Bmatrix} l \\ ij \end{Bmatrix}$ are not tensors, this notation is introduced in keeping with the summation convention which generally applies to two indices, one a superscript and the other a subscript. Thus all but one of the indices of the Christoffel symbols are regarded as subscripts. The exception is the l in the symbol of the second kind, which is treated as a superscript.

The definitions show that both symbols are symmetric with respect to the indices i and j. Inner multiplication of (20.2) by g_{lm} yields

$$(20.3) \qquad [ij, m] = g_{lm} \begin{Bmatrix} l \\ ij \end{Bmatrix}.$$

It follows immediately from (20.1) that

$$(20.4) \qquad \frac{\partial g_{ik}}{\partial x^j} = [ij, k] + [kj, i].$$

We now wish to express the derivatives of g^{ik} in terms of the Christoffel symbols, and so we differentiate equation (16.2) with respect to x^l, and obtain

$$\frac{\partial g^{ik}}{\partial x^l} g_{ij} + \frac{\partial g_{ij}}{\partial x^l} g^{ik} = 0.$$

Inner multiplication by g^{jm} gives us

$$\frac{\partial g^{mk}}{\partial x^l} + g^{jm} g^{ik} \frac{\partial g_{ij}}{\partial x^l} = 0.$$

Substituting from (20.4) and (20.2), we finally obtain

$$(20.5) \qquad \frac{\partial g^{mk}}{\partial x^l} = - g^{mi} \begin{Bmatrix} k \\ il \end{Bmatrix} - g^{ki} \begin{Bmatrix} m \\ il \end{Bmatrix}.$$

We shall now deduce a useful expression for $\begin{Bmatrix} i \\ ij \end{Bmatrix}$. Differentiate the determinant $g = | g_{ij} |$, remembering that $g^{lm}g$ is the cofactor of g_{lm} in this determinant, and obtain

$$\frac{\partial g}{\partial x^j} = g^{lm} g \frac{\partial g_{lm}}{\partial x^j}.$$

From (20.1) and (20.2) and the symmetry of g_{ij} we have

$$\begin{Bmatrix} i \\ ij \end{Bmatrix} = \frac{1}{2} g^{im} \left(\frac{\partial g_{im}}{\partial x^j} + \frac{\partial g_{jm}}{\partial x^i} - \frac{\partial g_{ij}}{\partial x^m} \right) = \frac{1}{2} g^{im} \frac{\partial g_{im}}{\partial x^j}.$$

Thus

$$\begin{Bmatrix} i \\ ij \end{Bmatrix} = \frac{1}{2} \frac{1}{g} \frac{\partial g}{\partial x^j} = \frac{\partial}{\partial x^j} \{\log \sqrt{g}\}. \tag{20.6}$$

Since g is not an invariant, it does not follow that $\begin{Bmatrix} i \\ ij \end{Bmatrix}$ is a covariant vector. If g is negative, equation (20.6) should be altered to $\begin{Bmatrix} i \\ ij \end{Bmatrix} = \frac{\partial}{\partial x^j} \{\log \sqrt{-g}\}$.

Ex. 1. Calculate the Christoffel symbols corresponding to the metrics

(a) $ds^2 = (dx^1)^2 + (x^1)^2(dx^2)^2 + (x^1)^2 \sin^2 x^2 (dx^3)^2$.

(b) $ds^2 = (dx^1)^2 + G(x^1, x^2)(dx^2)^2$, where G is a function of x^1 and x^2.

Ex. 2. If the metric of a V_N is such that $g_{ij} = 0$ for $i \neq j$, show that

$$\begin{Bmatrix} i \\ jk \end{Bmatrix} = 0; \quad \begin{Bmatrix} i \\ jj \end{Bmatrix} = -\frac{1}{2g_{ii}} \frac{\partial g_{jj}}{\partial x^i};$$

$$\begin{Bmatrix} i \\ ij \end{Bmatrix} = \frac{\partial}{\partial x^j} \{\log \sqrt{g_{ii}}\}; \quad \begin{Bmatrix} i \\ ii \end{Bmatrix} = \frac{\partial}{\partial x^i} \{\log \sqrt{g_{ii}}\},$$

where i, j and k are not equal, and the summation convention does not apply.

§ 21. Transformation law of Christoffel symbols

The fundamental tensor g_{ij}, being covariant, transforms according to the equation

$$\bar{g}_{lm} = \frac{\partial x^i}{\partial \bar{x}^l} \frac{\partial x^j}{\partial \bar{x}^m} g_{ij}. \tag{21.1}$$

On differentiating this equation with respect to $\bar{x}^n$, we have

$$\frac{\partial \bar{g}_{lm}}{\partial \bar{x}^n} = \frac{\partial x^i}{\partial \bar{x}^l} \frac{\partial x^j}{\partial \bar{x}^m} \frac{\partial g_{ij}}{\partial x^k} \frac{\partial x^k}{\partial \bar{x}^n} + \frac{\partial^2 x^i}{\partial \bar{x}^l \partial \bar{x}^n} \frac{\partial x^j}{\partial \bar{x}^m} g_{ij} + \frac{\partial x^i}{\partial \bar{x}^l} \frac{\partial^2 x^j}{\partial \bar{x}^m \partial \bar{x}^n} g_{ij}.$$

We subtract this equation from the sum of the two similar equations obtained by cyclic interchange of the indices l, m and n, and divide by two. Then with appropriate changes of dummy indices we have

$$(21.2)\quad \overline{[lm, n]} = [ij, k]\frac{\partial x^i}{\partial \bar{x}^l}\frac{\partial x^j}{\partial \bar{x}^m}\frac{\partial x^k}{\partial \bar{x}^n} + g_{ij}\frac{\partial x^i}{\partial \bar{x}^n}\frac{\partial^2 x^j}{\partial \bar{x}^l \partial \bar{x}^m},$$

where the bar over the Christoffel symbol indicates that it is calculated in the coordinate system $\bar{x}^i$ with respect to its fundamental tensor $\bar{g}_{ij}$. Now the transformation law of the contravariant fundamental tensor is

$$(21.3)\qquad \bar{g}^{np} = g^{rs}\frac{\partial \bar{x}^n}{\partial x^r}\frac{\partial \bar{x}^p}{\partial x^s}$$

Inner multiplication of both sides of equations (21.2) by the corresponding sides of (21.3) yields on reduction the relation

$$(21.4)\quad \overline{\begin{Bmatrix} p \\ lm \end{Bmatrix}} = \begin{Bmatrix} s \\ ij \end{Bmatrix}\frac{\partial \bar{x}^p}{\partial x^s}\frac{\partial x^i}{\partial \bar{x}^l}\frac{\partial x^j}{\partial \bar{x}^m} + \frac{\partial \bar{x}^p}{\partial x^j}\frac{\partial^2 x^j}{\partial \bar{x}^l \partial \bar{x}^m}.$$

Equations (21.2) and (21.4) comprise the transformation laws of the Christoffel symbols, and clearly indicate that they are not tensors. However, in the very special case of *linear* transformations of coordinates, $\partial^2 x^j/\partial \bar{x}^l \partial \bar{x}^m = 0$ and the symbols then transform like tensors. Now inner multiplication of (21.4) by $\partial x^r/\partial \bar{x}^p$ yields

$$(21.5)\qquad \frac{\partial^2 x^r}{\partial \bar{x}^l \partial \bar{x}^m} = \overline{\begin{Bmatrix} p \\ lm \end{Bmatrix}}\frac{\partial x^r}{\partial \bar{x}^p} - \begin{Bmatrix} r \\ ij \end{Bmatrix}\frac{\partial x^i}{\partial \bar{x}^l}\frac{\partial x^j}{\partial \bar{x}^m}.$$

This important equation expresses the second partial derivatives of the x^r with respect to the $\bar{x}^s$ in terms of the first derivatives and Christoffel symbols of the second kind.

Ex. Prove that the transformations of Christoffel symbols form a group.

§ 22. Covariant differentiation of vectors

Let us investigate the tensor character, if any, of the partial derivatives of a contravariant vector. We start by differentiating the transformation law

$$A^k = \bar{A}^i \frac{\partial x^k}{\partial \bar{x}^i} \tag{22.1}$$

with respect to x^j, and obtain

$$\frac{\partial A^k}{\partial x^j} = \frac{\partial \bar{A}^i}{\partial \bar{x}^n} \frac{\partial \bar{x}^n}{\partial x^j} \frac{\partial x^k}{\partial \bar{x}^i} + \bar{A}^i \frac{\partial^2 x^k}{\partial \bar{x}^i \partial \bar{x}^n} \frac{\partial \bar{x}^n}{\partial x^j}.$$

The presence of the last term on the right-hand side of this equation shows that the partial derivatives $\partial A^k/\partial x^j$ do not form a tensor. To obtain a tensor involving the partial derivatives we eliminate the partial derivatives of the second order by means of (21.5) and this gives us

$$\frac{\partial A^k}{\partial x^j} = \frac{\partial \bar{A}^i}{\partial \bar{x}^n} \frac{\partial \bar{x}^n}{\partial x^j} \frac{\partial x^k}{\partial \bar{x}^i} + \bar{A}^i \frac{\partial \bar{x}^n}{\partial x^j} \left[\overline{\begin{Bmatrix} p \\ in \end{Bmatrix}} \frac{\partial x^k}{\partial \bar{x}^p} - \begin{Bmatrix} k \\ rs \end{Bmatrix} \frac{\partial x^r}{\partial \bar{x}^i} \frac{\partial x^s}{\partial \bar{x}^n} \right].$$

In virtue of (22.1) and by appropriate changes of dummy indices, this equation reduces to

$$\frac{\partial A^k}{\partial x^j} + \begin{Bmatrix} k \\ rj \end{Bmatrix} A^r = \left[\frac{\partial \bar{A}^i}{\partial \bar{x}^n} + \overline{\begin{Bmatrix} i \\ rn \end{Bmatrix}} \bar{A}^r \right] \frac{\partial \bar{x}^n}{\partial x^j} \frac{\partial x^k}{\partial \bar{x}^i}.$$

Introduce the comma notation

$$A^k_{,j} \equiv \frac{\partial A^k}{\partial x^j} + \begin{Bmatrix} k \\ rj \end{Bmatrix} A^r, \tag{22.2}$$

so that the above equation can be written

$$A^k_{,j} = \bar{A}^i_{,n} \frac{\partial \bar{x}^n}{\partial x^j} \frac{\partial x^k}{\partial \bar{x}^i}.$$

Hence from (8.3) it is obvious that $A^k_{,j}$ is a mixed tensor of the second order and it is called the **covariant derivative** of A^k with respect to x^j.

In order to set up the corresponding entity for covariant vectors, we may conveniently start by differentiating the transformation law

$$\bar{A}_i = A_j \frac{\partial x^j}{\partial \bar{x}^i} \tag{22.3}$$

with respect to $\bar{x}^l$. This yields

$$\frac{\partial \bar{A}_i}{\partial \bar{x}^l} = \frac{\partial A_j}{\partial x^k} \frac{\partial x^k}{\partial \bar{x}^l} \frac{\partial x^j}{\partial \bar{x}^i} + A_j \frac{\partial^2 x^j}{\partial \bar{x}^i \partial \bar{x}^l}.$$

Again, by (21.5), we eliminate the partial derivatives of the second order. Further we change dummy indices as required and we have on substitution from (22.3), that

$$\frac{\partial \bar{A}_i}{\partial \bar{x}^l} - \overline{\left\{ {m \atop il} \right\}} \bar{A}_m = \left[\frac{\partial A_j}{\partial x^n} - \left\{ {r \atop jn} \right\} A_r \right] \frac{\partial x^j}{\partial \bar{x}^i} \frac{\partial x^n}{\partial \bar{x}^l}.$$

The comma notation

$$A_{j,n} \equiv \frac{\partial A_j}{\partial x^n} - \left\{ {r \atop jn} \right\} A_r \tag{22.4}$$

is now introduced, and the above equation can be written

$$\bar{A}_{i,l} = A_{j,n} \frac{\partial x^j}{\partial \bar{x}^i} \frac{\partial x^n}{\partial \bar{x}^l},$$

which shows that $A_{j,n}$ is a covariant tensor of the second order, called the **covariant derivative** of A_j with respect to x^n.

In a Euclidean space of N dimensions, referred to rectangular cartesian coordinates, the components of the fundamental tensor g_{ij} are zero except $g_{11} = g_{22} = \ldots = g_{NN} = 1$. Thus all the Christoffel symbols are zero and so covariant differentiation reduces to the familiar partial differentiation. It is well to observe that the Christoffel symbols do *not* all vanish in a Euclidean space referred, for example, to spherical polar coordinates.

We can construct the invariant $A^j_{,j}$ by contraction.

Applying (20.6) we obtain

$$A^j_{,j} = \frac{\partial A^j}{\partial x^j} + \begin{Bmatrix} j \\ rj \end{Bmatrix} A^r = \frac{\partial A^j}{\partial x^j} + A^r \frac{\partial}{\partial x^r} \{\log \sqrt{g}\}.$$

That is

$$A^j_{,j} = \frac{1}{\sqrt{g}} \frac{\partial}{\partial x^r} \{\sqrt{g} A^r\}. \tag{22.5}$$

This invariant is called the **divergence** of the contravariant vector A^i and is often denoted by div A^i. The divergence of a covariant vector A_i is defined by

$$\operatorname{div} A_i = g^{jk} A_{j,k}. \tag{22.6}$$

The partial derivatives of an invariant form the components of a covariant vector. We extend the definition of covariant differentiation to invariants by calling the familiar partial derivative the covariant derivative. That is, from the invariant I we form by definition

$$I_{,i} \equiv \frac{\partial I}{\partial x^i}.$$

Since $I_{,i}$ is a covariant vector, we find from (22.4) that its covariant derivative with respect to x^j is given by

$$(I_{,i})_{,j} = \frac{\partial^2 I}{\partial x^i \partial x^j} - \begin{Bmatrix} r \\ ij \end{Bmatrix} \frac{\partial I}{\partial x^r}. \tag{22.7}$$

Thus $(I_{,i})_{,j} = (I_{,j})_{,i}$. That is, the covariant differentiation of invariants is commutative. In section 31, we shall see that the covariant differentiation of vectors is not in general commutative.

We can form the divergence of the covariant vector $I_{,i}$; this is called the **Laplacian** of I and written $\nabla^2 I$. Then

$$\nabla^2 I = g^{jk}(I_{,j})_{,k} = g^{jk} \left(\frac{\partial^2 I}{\partial x^j \partial x^k} - \begin{Bmatrix} r \\ jk \end{Bmatrix} \frac{\partial I}{\partial x^r} \right).$$

Ex. 1. Show that $A_{j,n} - A_{n,j} = \dfrac{\partial A_j}{\partial x^n} - \dfrac{\partial A_n}{\partial x^j}$.

Ex. 2. Prove that $\operatorname{div} A_j = \dfrac{1}{\sqrt{g}} \dfrac{\partial}{\partial x^r} \{\sqrt{g}\, g^{rk} A_k\} = \operatorname{div} A^j.$

Ex. 3. Evaluate div A^j and $\nabla^2 I$ in (a) cylindrical polar, (b) spherical polar coordinates.

§ 23. Covariant differentiation of tensors

In the last section we constructed tensors containing the partial derivatives of vectors. Can we now extend the process of covariant differentiation to tensors? For our typical tensor we shall choose A^i_j. There will be no loss in generality involved as this tensor has one contravariant and one covariant index. Inner multiplication of its transformation law (8.3) by $\partial x^m/\partial \bar{x}^i$ yields

$$\bar{A}^i_j \frac{\partial x^m}{\partial \bar{x}^i} = A^m_l \frac{\partial x^l}{\partial \bar{x}^j}. \tag{23.1}$$

We differentiate with respect to $\bar{x}^k$, then eliminate the second order partial derivatives by means of (21.5) and obtain

$$\frac{\partial \bar{A}^i_j}{\partial \bar{x}^k} \frac{\partial x^m}{\partial \bar{x}^i} + \bar{A}^i_j \left[\overline{\begin{Bmatrix} p \\ ik \end{Bmatrix}} \frac{\partial x^m}{\partial \bar{x}^p} - \begin{Bmatrix} m \\ rs \end{Bmatrix} \frac{\partial x^r}{\partial \bar{x}^i} \frac{\partial x^s}{\partial \bar{x}^k} \right]$$
$$= \frac{\partial A^m_l}{\partial x^t} \frac{\partial x^t}{\partial \bar{x}^k} \frac{\partial x^l}{\partial \bar{x}^j} + A^m_l \left[\overline{\begin{Bmatrix} p \\ jk \end{Bmatrix}} \frac{\partial x^l}{\partial \bar{x}^p} - \begin{Bmatrix} l \\ rs \end{Bmatrix} \frac{\partial x^r}{\partial \bar{x}^j} \frac{\partial x^s}{\partial \bar{x}^k} \right].$$

Applying (23.1) to this and changing appropriate dummy indices, this equation becomes

$$\left[\frac{\partial \bar{A}^i_j}{\partial \bar{x}^k} + \bar{A}^n_j \overline{\begin{Bmatrix} i \\ nk \end{Bmatrix}} - \bar{A}^i_p \overline{\begin{Bmatrix} p \\ jk \end{Bmatrix}} \right] \frac{\partial x^m}{\partial \bar{x}^i}$$
$$= \left[\frac{\partial A^m_l}{\partial x^t} + A^r_l \begin{Bmatrix} m \\ rt \end{Bmatrix} - A^m_r \begin{Bmatrix} r \\ lt \end{Bmatrix} \right] \frac{\partial x^t}{\partial \bar{x}^k} \frac{\partial x^l}{\partial \bar{x}^j}.$$

Introduce the comma notation

$$A^m_{l,t} \equiv \frac{\partial A^m_l}{\partial x^t} + \begin{Bmatrix} m \\ rt \end{Bmatrix} A^r_l - \begin{Bmatrix} r \\ lt \end{Bmatrix} A^m_r, \tag{23.2}$$

and the above equation takes the form

$$\bar{A}^i_{j,k}\frac{\partial x^m}{\partial \bar{x}^i} = A^m_{l,t}\frac{\partial x^t}{\partial \bar{x}^k}\frac{\partial x^l}{\partial \bar{x}^j}.$$

On inner multiplication by $\partial \bar{x}^r/\partial x^m$ we have

$$\bar{A}^r_{j,k} = A^m_{l,t}\frac{\partial \bar{x}^r}{\partial x^m}\frac{\partial x^l}{\partial \bar{x}^j}\frac{\partial x^t}{\partial \bar{x}^k}.$$

Hence $A^m_{l,t}$ is a tensor of the third order of the type indicated by its indices. This tensor is called the **covariant derivative** of A^m_l with respect to x^t.

An examination of equation (23.2) shows that the covariant derivative of A^m_l contains three terms. They are (1) the partial derivative, (2) a term with positive sign similar to that contained in the covariant derivative of a contravariant vector and (3) a term with negative sign similar to that contained in the covariant derivative of a covariant vector. This suggests that the expression

$$\text{(23.3)}\quad \begin{aligned} A^{u_1..u_s}_{r_1..r_p,n} \equiv \frac{\partial A^{u_1..u_s}_{r_1..r_p}}{\partial x^n} &+ \sum_{\alpha=1}^{s}\begin{Bmatrix} u_\alpha \\ kn \end{Bmatrix} A^{u_1..u_{\alpha-1}ku_{\alpha+1}..u_s}_{r_1..r_p} \\ &- \sum_{\beta=1}^{p}\begin{Bmatrix} l \\ r_\beta n \end{Bmatrix} A^{u_1..u_s}_{r_1..r_{\beta-1}lr_{\beta+1}..r_p} \end{aligned}$$

is a tensor, called the **covariant derivative** of $A^{u_1..u_s}_{r_1..r_p}$ with respect to x^n. The proof which we worked out for the mixed tensor of the second order applies also to (23.3). We will omit its tedious details and supply an alternative proof in section 30.

On referring to equations (20.3), (20.4), (20.5) and (23.3) we deduce that

$$\text{(23.4)}\quad g_{ij,k} = \frac{\partial g_{ij}}{\partial x^k} - \begin{Bmatrix} l \\ ik \end{Bmatrix} g_{lj} - \begin{Bmatrix} l \\ jk \end{Bmatrix} g_{il} = 0,$$

$$\text{(23.5)}\quad g^{ij}_{,k} = \frac{\partial g^{ij}}{\partial x^k} + \begin{Bmatrix} i \\ lk \end{Bmatrix} g^{lj} + \begin{Bmatrix} j \\ lk \end{Bmatrix} g^{il} = 0$$

and

$$\delta^i_{j,k} = \frac{\partial \delta^i_j}{\partial x^k} + \begin{Bmatrix} i \\ lk \end{Bmatrix} \delta^l_j - \begin{Bmatrix} l \\ jk \end{Bmatrix} \delta^i_l = 0. \tag{23.6}$$

The covariant derivatives of covariant derivatives are again tensors. We indicate these covariant derivatives of the second order by adding another index without a comma. For example $A_{i,jk}$ is the covariant derivative of $A_{i,j}$ with respect to x^k.

Ex. 1. If $A_{ij} = B_{i,j} - B_{j,i}$, prove that $A_{ij,k} + A_{jk,i} + A_{ki,j} = 0$.
Ex. 2. By means of (23.5), show that div A^i = div A_i.
Ex. 3. If A^{ijk} is a skew-symmetric tensor, show that

$$\frac{1}{\sqrt{g}} \frac{\partial}{\partial x^k} (\sqrt{g}\, A^{ijk}) \text{ is a tensor.}$$

§ 24. Laws of covariant differentiation

Covariant derivatives obey the following laws:-

(1) the covariant derivative of the sum (or difference) of two tensors is the sum (or difference) of their covariant derivatives. This law is an immediate deduction from (23.3).

(2) the covariant derivative of an outer (or inner) product of two tensors is equal to the sum of the two terms obtained by outer (or inner) multiplication of each tensor with the covariant derivative of the other tensor. Consider as an illustration

$$(A_{ij}B^l)_{,m} = \frac{\partial}{\partial x^m}(A_{ij}B^l) + A_{ij}\begin{Bmatrix} l \\ mk \end{Bmatrix} B^k - B^l\left[\begin{Bmatrix} k \\ im \end{Bmatrix} A_{kj} + \begin{Bmatrix} k \\ jm \end{Bmatrix} A_{ik}\right]$$
$$= A_{ij,m}B^l + A_{ij}B^l_{,m}.$$

This type of proof is quite general and will apply to any case of outer multiplication of two tensors. (Another proof is provided in section 28). Contraction of l and j gives us

$$(A_{ij}B^j)_{,m} = A_{ij,m}B^j + A_{ij}B^j_{,m},$$

which makes it clear that the rule also applies to inner products.

(3) the tensors g_{ij}, g^{ij} and δ^i_j are constants with respect to covariant differentiation. This is merely another way of stating equations (23.4), (23.5) and (23.6). For example

$$(g^{ij}A_{il})_{,m} = g^{ij}_{,m}A_{il} + g^{ij}A_{il,m} = g^{ij}A_{il,m}.$$

Ex. If I and J are invariants, show that

$$\text{div } (JI_{,i}) = J\nabla^2 I + g^{ij}I_{,i}J_{,j}.$$

§ 25. Intrinsic derivatives

Consider the tensor $A^{u_1 \cdots u_s}_{r_1 \cdots r_p}$ whose components are functions of t along a curve $x^i = x^i(t)$. The **intrinsic derivative** is defined by

$$\frac{\delta A^{u_1 \cdots u_s}_{r_1 \cdots r_p}}{\delta t} \equiv A^{u_1 \cdots u_s}_{r_1 \cdots r_p, k} \frac{dx^k}{dt}. \tag{25.1}$$

Accordingly, the intrinsic derivative is a tensor of the *same* order and type as the original tensor.

Corresponding to the invariant I, we have

$$\frac{\delta I}{\delta t} = I_{,k} \frac{dx^k}{dt} = \frac{\partial I}{\partial x^k} \frac{dx^k}{dt} = \frac{dI}{dt}.$$

That is, the intrinsic derivative of an invariant coincides with its total derivative.

Intrinsic derivatives of higher order are easily defined. For example,

$$\frac{\delta^2 A^i_j}{\delta t^2} = \frac{\delta}{\delta t}\left(\frac{\delta A^i_j}{\delta t}\right) = \left(A^i_{j,k} \frac{dx^k}{dt}\right)_{,l} \frac{dx^l}{dt}.$$

In general, intrinsic differentiation is not commutative.

From (22.2), (22.4), (23.4), (23.5) and (23.6) we calculate that

$$\frac{\delta A^k}{\delta t} \equiv \frac{dA^k}{dt} + \begin{Bmatrix} k \\ rj \end{Bmatrix} A^r \frac{dx^j}{dt}, \tag{25.2}$$

$$\frac{\delta A_k}{\delta t} \equiv \frac{dA_k}{dt} - \begin{Bmatrix} r \\ kj \end{Bmatrix} A_r \frac{dx^j}{dt}, \tag{25.3}$$

and

$$\frac{\delta g_{ij}}{\delta t} = \frac{\delta g^{ij}}{\delta t} = \frac{\delta \delta^i_j}{\delta t} = 0. \tag{25.4}$$

From the definition of intrinsic derivatives, it follows that they obey the same three laws which apply to covariant derivatives.

Ex. Show that $\dfrac{\delta}{\delta t}\left(\dfrac{dx^i}{dt}\right) \equiv \dfrac{d^2x^i}{dt^2} + \begin{Bmatrix} i \\ jk \end{Bmatrix} \dfrac{dx^j}{dt}\dfrac{dx^k}{dt}$.

Solutions

p. 28 Ex. 1. The only non-zero Christoffel symbols of the second kind are

(a) $\begin{Bmatrix} 1 \\ 22 \end{Bmatrix} = -x^1$, $\begin{Bmatrix} 1 \\ 33 \end{Bmatrix} = -x^1 \sin^2 x^2$, $\begin{Bmatrix} 2 \\ 12 \end{Bmatrix} = 1/x^1$,

$\begin{Bmatrix} 2 \\ 33 \end{Bmatrix} = -\sin x^2 \cos x^2$, $\begin{Bmatrix} 3 \\ 13 \end{Bmatrix} = 1/x^1$, $\begin{Bmatrix} 3 \\ 23 \end{Bmatrix} = \cot x^2$;

(b) $\begin{Bmatrix} 1 \\ 22 \end{Bmatrix} = -\dfrac{1}{2}\dfrac{\partial G}{\partial x^1}$, $\begin{Bmatrix} 2 \\ 12 \end{Bmatrix} = \dfrac{1}{2G}\dfrac{\partial G}{\partial x^1}$, $\begin{Bmatrix} 2 \\ 22 \end{Bmatrix} = \dfrac{1}{2G}\dfrac{\partial G}{\partial x^2}$.

p. 33 Ex. 3.

(a) $$\operatorname{div} A^i = \frac{\partial A^1}{\partial x^1} + \frac{\partial A^2}{\partial x^2} + \frac{\partial A^3}{\partial x^3} + \frac{1}{x^1}A^1,$$

$$= \frac{\partial A_1}{\partial x^1} + \frac{1}{(x^1)^2}\frac{\partial A_2}{\partial x^2} + \frac{\partial A_3}{\partial x^3} + \frac{1}{x^1}A_1;$$

$$\nabla^2 I = \frac{\partial^2 I}{(\partial x^1)^2} + \frac{1}{(x^1)^2}\frac{\partial^2 I}{(\partial x^2)^2} + \frac{\partial^2 I}{(\partial x^3)^2} + \frac{1}{x^1}\frac{\partial I}{\partial x^1}$$

(b) $$\operatorname{div} A^i = \frac{\partial A^1}{\partial x^1} + \frac{\partial A^2}{\partial x^2} + \frac{\partial A^3}{\partial x^3} + \frac{2}{x^1}A^1 + \cot x^2 A^2,$$

$$= \frac{\partial A_1}{\partial x^1} + \frac{1}{(x^1)^2}\frac{\partial A_2}{\partial x^2} + \frac{1}{(x^1)^2 \sin^2 x^2}\frac{\partial A_3}{\partial x^3} + \frac{2}{x^1}A_1 + \frac{\cot x^2}{(x^1)^2}A_2;$$

$$\nabla^2 I = \frac{\partial^2 I}{(\partial x^1)^2} + \frac{1}{(x^1)^2}\frac{\partial^2 I}{(\partial x^2)^2} + \frac{1}{(x^1)^2 \sin^2 x^2}\frac{\partial^2 I}{(\partial x^3)^2} + \frac{2}{x^1}\frac{\partial I}{\partial x^1} + \frac{\cot x^2}{(x^1)^2}\frac{\partial I}{\partial x^2}.$$

The corresponding results in Rutherford's Vector Methods, pp. 72—73, differ slightly because they are expressed in terms of the physical components of the above vector (see section 62).

CHAPTER IV

GEODESICS – PARALLELISM

§ 26. Geodesics

In Euclidean three-dimensional space, a straight line is the path of shortest distance between two points, and it is our aim to generalise this fundamental concept to Riemannian spaces. Let C be the curve $x^i = x^i(t)$ with parameter t joining two fixed points P_0 and P_1, whose parameters are t_0 and t_1 respectively. Then the distance s along the curve between P_0 and P_1 is given by

$$s = \int_{t_0}^{t_1} \sqrt{e g_{ij} \frac{dx^i}{dt} \frac{dx^j}{dt}}\, dt. \tag{26.1}$$

Consider all the curves passing through the two fixed points P_0 and P_1. Any of these curves, for which the distance P_0P_1 measured along the curve is stationary, is called a **geodesic.** We could obtain the differential equations of the geodesics by applying Euler's equations, a well-known result in the calculus of variations, to equations (26.1). However we shall find it more instructive to appeal to first principles.

Let us choose a small *arbitrary* vector δx^i varying continuously along C. Then the equations $\bar{x}^i = x^i + \delta x^i$ define a neighbouring curve $\bar{C}$ to C. Further let us impose the conditions that $\delta x^i = 0$ at P_0 and P_1. This means that the curve $\bar{C}$ always joins P_0 to P_1. The distance $\bar{s}$ from P_0 to P_1 along $\bar{C}$ is given by

$$\bar{s} = \int_{t_0}^{t_1} \sqrt{e g_{ij}(\bar{x}) \frac{d\bar{x}^i}{dt} \frac{d\bar{x}^j}{dt}}\, dt,$$

where the $g_{ij}(\bar{x})$ are now functions of $\bar{x}^i$. We have

$$g_{ij}(\bar{x})\frac{d\bar{x}^i}{dt}\frac{d\bar{x}^j}{dt}=\left(g_{ij}+\frac{\partial g_{ij}}{\partial x^k}\delta x^k\right)\left(\frac{dx^i}{dt}+\frac{d(\delta x^i)}{dt}\right)\left(\frac{dx^j}{dt}+\frac{d(\delta x^j)}{dt}\right)$$

$$=g_{ij}\frac{dx^i}{dt}\frac{dx^j}{dt}+2g_{ij}\frac{dx^i}{dt}\frac{d(\delta x^j)}{dt}+\frac{\partial g_{ij}}{\partial x^k}\delta x^k\frac{dx^i}{dt}\frac{dx^j}{dt},$$

where terms of higher order than the first are neglected. Thus

$$\sqrt{eg_{ij}(\bar{x})\frac{d\bar{x}^i}{dt}\frac{d\bar{x}^j}{dt}}=$$

$$\sqrt{eg_{ij}\frac{dx^i}{dt}\frac{dx^j}{dt}}\left[1+\frac{g_{ij}\dfrac{dx^i}{dt}\dfrac{d(\delta x^j)}{dt}+\dfrac{1}{2}\dfrac{\partial g_{ij}}{\partial x^k}\delta x^k\dfrac{dx^i}{dt}\dfrac{dx^j}{dt}}{g_{ij}\dfrac{dx^i}{dt}\dfrac{dx^j}{dt}}\right].$$

Consequently the variation in length δs from the curve C to the curve $\bar{C}$ is given by

$$\delta s=\bar{s}-s$$

$$=\int_{t_0}^{t_1}\frac{g_{ij}\dfrac{dx^i}{dt}\dfrac{d(\delta x^j)}{dt}+\dfrac{1}{2}\dfrac{\partial g_{ij}}{\partial x^k}\delta x^k\dfrac{dx^i}{dt}\dfrac{dx^j}{dt}}{\sqrt{eg_{ij}\dfrac{dx^i}{dt}\dfrac{dx^j}{dt}}}\,dt.$$

We now simplify this equation by choosing the arc-distance s along C as parameter and obtain

$$\delta s=\int_{s_0}^{s_1}\left[g_{ij}\frac{dx^i}{ds}\frac{d(\delta x^j)}{ds}+\frac{1}{2}\frac{\partial g_{ij}}{\partial x^k}\delta x^k\frac{dx^i}{ds}\frac{dx^j}{ds}\right]ds,$$

where s_0 and s_1 are the values of s corresponding to the points P_0 and P_1 respectively. Integration by parts yields

$$\delta s=\left[g_{ij}\frac{dx^i}{ds}\delta x^j\right]_{s_0}^{s_1}-\int_{s_0}^{s_1}\delta x^j\left[\frac{d}{ds}\left(g_{ij}\frac{dx^i}{ds}\right)-\frac{1}{2}\frac{\partial g_{ik}}{\partial x^j}\frac{dx^i}{ds}\frac{dx^k}{ds}\right]ds.$$

The integrated portion vanishes since δx^j is zero at P_0

and P_1. Also we have

$$\frac{d}{ds}\left(g_{ij}\frac{dx^i}{ds}\right) = g_{ij}\frac{d^2x^i}{ds^2} + \frac{\partial g_{ij}}{\partial x^k}\frac{dx^i}{ds}\frac{dx^k}{ds}$$

$$= g_{ij}\frac{d^2x^i}{ds^2} + \frac{1}{2}\frac{\partial g_{ij}}{\partial x^k}\frac{dx^i}{ds}\frac{dx^k}{ds} + \frac{1}{2}\frac{\partial g_{kj}}{\partial x^i}\frac{dx^i}{ds}\frac{dx^k}{ds}.$$

It follows that

$$(26.2) \quad \delta s = -\int_{s_0}^{s_1} \delta x^j \left[g_{ij}\frac{d^2x^i}{ds^2} + [ik, j]\frac{dx^i}{ds}\frac{dx^k}{ds}\right] ds.$$

The variations δx^j are *arbitrary*, thus the necessary and sufficient conditions that the curve C be a geodesic are

$$(26.3) \qquad g_{ij}\frac{d^2x^i}{ds^2} + [ik, j]\frac{dx^i}{ds}\frac{dx^k}{ds} = 0.$$

Inner multiplication by g^{jl} yields the contravariant form

$$(26.4) \qquad \frac{\delta}{\delta s}\left(\frac{dx^l}{ds}\right) \equiv \frac{d^2x^l}{ds^2} + \left\{\begin{matrix} l \\ ik \end{matrix}\right\}\frac{dx^i}{ds}\frac{dx^k}{ds} = 0.$$

Either set of equations (26.3) or (26.4) are the differential equations of a geodesic. They constitute N differential equations of the second order. The theory of differential equations states that a solution $x^i = x^i(s)$ is determined uniquely if the initial values of x^i and dx^i/ds are given at any point. Geometrically this means that there is a unique geodesic with given direction at any point of the space. We defined the geodesic in terms of the curve passing through two points, but this geodesic may not be unique unless the two points are sufficiently close to one another. The problem of uniqueness now involves topological properties of the space V_N. For example, there is a unique geodesic passing through two points on a sphere, except when the two points are at the ends of a diameter. In this latter case, all great circles passing through the two points are geodesics.

For a Euclidean space, referred to rectangular cartesian coordinates, the Christoffel symbols are zero. Hence the geodesics are given by $d^2x^l/ds^2 = 0$, whose solution is $x^l = A^l s + B^l$, where A^l and B^l are constant vectors. That is, the geodesics are straight lines.

§ 27. Null-geodesics

Equations (15.3) state that

$$g_{ij}\frac{dx^i}{ds}\frac{dx^j}{ds} = e \tag{27.1}$$

along any portion of a curve which is not null. On differentiation we obtain

$$\frac{d}{ds}\left(g_{ij}\frac{dx^i}{ds}\frac{dx^j}{ds}\right) = \frac{\delta}{\delta s}\left(g_{ij}\frac{dx^i}{ds}\frac{dx^j}{ds}\right) = 2g_{ij}\frac{dx^i}{ds}\frac{\delta}{\delta s}\left(\frac{dx^j}{ds}\right).$$

It follows from (26.4) that the invariant $\frac{d}{ds}\left(g_{ij}\frac{dx^i}{ds}\frac{dx^j}{ds}\right)$ is zero at all points on a geodesic. Thus the indicator e cannot change abruptly along a geodesic, and so if the tangent vector is not null at any one point, it cannot be null at any other point on the geodesic. On the other hand, if the initial direction is null, then the curve is null and it is of course impossible to introduce the arc-distance as parameter. Instead we now say that a null curve $x^i = x^i(t)$ which is a solution of the equations

$$\frac{d^2x^l}{dt^2} + \begin{Bmatrix} l \\ ik \end{Bmatrix}\frac{dx^i}{dt}\frac{dx^k}{dt} = 0 \tag{27.2}$$

is a **null-geodesic.**

The null-geodesics in the V_4 with line-element (14.2) satisfy the equations $\frac{d^2x^l}{dt^2} = 0$. Therefore the null curves given by (15.2) do not satisfy these equations unless r, θ and ψ are constants. It follows that a null curve is not necessarily a null-geodesic.

§ 28. Geodesic coordinates

We shall now show that it is always possible to choose the coordinate system so that all the Christoffel symbols are zero at a particular point. Consider a general coordinate system x^i, whose values at a particular point P_0 are $x^i_{(0)}$, and introduce a new coordinate system $\bar{x}^i$ by the equations

$$(28.1) \qquad \bar{x}^i = x^i - x^i_{(0)} + \frac{1}{2} \begin{Bmatrix} i \\ mn \end{Bmatrix}_{(0)} (x^m - x^m_{(0)})(x^n - x^n_{(0)}).$$

The index (0) attached to any entity denotes its value at the point P_0. The brackets serve to emphasise that this index has no tensorial significance and that the summation convention does not apply to it. Differentiation with respect to x^j yields

$$(28.2) \qquad \frac{\partial \bar{x}^i}{\partial x^j} = \delta^i_j + \begin{Bmatrix} i \\ jn \end{Bmatrix}_{(0)} (x^n - x^n_{(0)}).$$

Hence $\left(\frac{\partial \bar{x}^i}{\partial x^j}\right)_{(0)} = \delta^i_j$. Accordingly the Jacobian determinant $\left| \left(\frac{\partial \bar{x}^i}{\partial x^j}\right)_{(0)} \right|$ is not zero, which shows us that the transformation (28.1) is permissible in the neighbourhood of P_0. On inner multiplication of (28.2) with $\partial x^j / \partial \bar{x}^k$, we obtain

$$\delta^i_k = \frac{\partial x^i}{\partial \bar{x}^k} + \begin{Bmatrix} i \\ jn \end{Bmatrix}_{(0)} (x^n - x^n_{(0)}) \frac{\partial x^j}{\partial \bar{x}^k}.$$

We differentiate this with respect to $\bar{x}^h$ and obtain

$$0 = \frac{\partial^2 x^i}{\partial \bar{x}^k \partial \bar{x}^h} + \begin{Bmatrix} i \\ jn \end{Bmatrix}_{(0)} \frac{\partial x^n}{\partial \bar{x}^h} \frac{\partial x^j}{\partial \bar{x}^k} + \begin{Bmatrix} i \\ jn \end{Bmatrix}_{(0)} (x^n - x^n_{(0)}) \frac{\partial^2 x^j}{\partial \bar{x}^k \partial \bar{x}^h}.$$

Thus at P_0

$$\left(\frac{\partial x^i}{\partial \bar{x}^k}\right)_{(0)} = \delta^i_k; \left(\frac{\partial^2 x^i}{\partial \bar{x}^k \partial \bar{x}^h}\right)_{(0)} = - \begin{Bmatrix} i \\ jn \end{Bmatrix}_{(0)} \delta^n_h \delta^j_k = - \begin{Bmatrix} i \\ kh \end{Bmatrix}_{(0)}.$$

We now substitute in (21.4) and derive

$$\overline{\begin{Bmatrix} p \\ lm \end{Bmatrix}}_{(0)} = \begin{Bmatrix} s \\ ij \end{Bmatrix}_{(0)} \delta^p_s \delta^i \delta^j_m - \delta^p_j \begin{Bmatrix} j \\ lm \end{Bmatrix}_{(0)}.$$

That is,

$$\overline{\begin{Bmatrix} p \\ lm \end{Bmatrix}}_{(0)} = 0.$$

Hence a particular system of coordinates, called **geodesic coordinates**, can always be chosen so that the Christoffel symbols are zero at any assigned point called the **pole.** The transformation (28.1) is not the only method of obtaining geodesic coordinates. However, we have shown that one such system exists and in this particular system $\bar{x}^i_{(0)} = 0$, which means that the pole is also the origin of coordinates.

The important property possessed by a geodesic coordinate system is this: the covariant derivatives reduce to the corresponding partial derivatives at the pole because at this point all the Christoffel symbols are zero. We mentioned earlier in section 9 the fact that if a tensor is zero in one coordinate system, it is zero in every coordinate system, since the transformation law of tensors is linear. The setting up of a tensor equation often involves heavy algebraic manipulations. Generally, the volume of work is reduced by first proving the equation with respect to a geodesic coordinate system at its pole. It follows that the equation is true for all coordinate systems at this point. Then if this point is general, the equation is true at all points of the V_N.

We shall illustrate this method by proving the multiplication law of section 24 which is satisfied by covariant derivatives. Consider the tensor

$$(A_{ij}B^j)_{,m} - A_{ij,m}B^j - A_{ij}B^j_{,m}.$$

Let us choose a geodesic coordinate system with pole at P_0, then the covariant derivatives reduce to the familiar partial derivatives at P_0. Since partial derivatives satisfy the multiplication law

$$\frac{\partial}{\partial u}(\varphi\psi) = \frac{\partial\varphi}{\partial u}\psi + \varphi\frac{\partial\psi}{\partial u},$$

the above tensor is zero at P_0 in the geodesic coordinate system. Thus it is zero at P_0 in every coordinate system. But P_0 is a general point, therefore the above tensor is zero at all points in the V_N. This establishes the truth of the multiplication law.

Ex. Show that at the pole P_0 of a geodesic coordinate system

$$A_{i,jk} = \frac{\partial^2 A_i}{\partial x^j \partial x^k} - A_l \frac{\partial}{\partial x^k} \begin{Bmatrix} l \\ ij \end{Bmatrix}.$$

§ 29. Parallelism

An important property of parallelism in a Euclidean space, which is referred to rectangular cartesian coordinates, is this: a parallel field of vectors A_i is obtained throughout a Euclidean space if the components A_i are constants. We can express this analytically either in the form $dA_i/dt = 0$ or by $\partial A_i/\partial x^j = 0$. Since the Christoffel symbols are zero, we can write these equations equivalently in the tensorial forms $\delta A_i/\delta t = 0$ or $A_{i,j} = 0$ respectively. This suggests two ways of generalising the concept of parallelism to a Riemannian space. We shall see in section 31 that the partial differential equations $A_{i,j} = 0$ are in general not consistent. Thus the second suggested generalisation is not a profitable one. Also the intrinsic derivative $\delta A_i/\delta t$ is only defined along a curve, hence, using the first suggestion, we can only define parallelism along a curve. Formally, the vectors A_i constitute a field of **parallel vectors** along the curve $x^i = x^i(t)$ if A_i is a solution of the differential equations

$$\frac{\delta A_i}{\delta t} \equiv \frac{dA_i}{dt} - \begin{Bmatrix} l \\ ik \end{Bmatrix} A_l \frac{dx^k}{dt} = 0. \tag{29.1}$$

These equations form a set of N differential equations of the first order, and consequently if the vector A_i is given at any one point on the curve, it is uniquely determined at all other points of the curve. We may also say that a field of parallel vectors is obtained from a given vector by parallel propagation along the curve. Since

$$\frac{\delta A^i}{\delta t} = \frac{\delta}{\delta t}(g^{ij}A_j) = g^{ij}\frac{\delta A_j}{\delta t}$$

we can write the condition for parallelism along a curve in the contravariant form

$$\frac{\delta A^i}{\delta t} \equiv \frac{dA^i}{dt} + \begin{Bmatrix} i \\ jk \end{Bmatrix} A^j \frac{dx^k}{dt} = 0. \tag{29.2}$$

We see from equations (26.4) that the unit tangent vectors form a field of parallel vectors along a geodesic.

The magnitude A of the vector A^i is given by $(A)^2 = e_{(A)}\, g_{ij} A^i A^j$. On differentiation we have

$$2A\frac{dA}{dt} = \frac{d}{dt}(e_{(A)}g_{ij}A^iA^j) = \frac{\delta}{\delta t}(e_{(A)}g_{ij}A^iA^j) = 2e_{(A)}g_{ij}\frac{\delta A^i}{\delta t}A^j.$$

This equation becomes $A(dA/dt) = 0$ if A^i forms a field of parallel vectors. We deduce that the magnitude of all vectors of a field of parallel vectors is constant.

The angle θ between the two vectors A^i and B^i is given by $AB \cos\theta = g_{ij}A^iB^j$. We obtain on differentiation

$$AB\frac{d}{dt}(\cos\theta) + \left(\frac{dA}{dt}B + A\frac{dB}{dt}\right)\cos\theta = g_{ij}\left(\frac{\delta A^i}{\delta t}B^j + A^i\frac{\delta B^j}{\delta t}\right).$$

When both A^i and B^i form fields of parallel vectors, this equation reduces to $d(\cos\theta)/dt = 0$, provided that neither of the vectors A^i and B^i is null. That is, the angle between two non-null vectors remains constant whilst both undergo parallel propagation along the same curve.

The vector obtained at Q by parallel propagation from P depends on the curve joining P to Q. Hence parallel propagation around a closed curve does not necessarily lead back to the initial vector. As an example consider the V_2 formed by the surface of a unit sphere. On choosing spherical polar coordinates $x^1 = \theta$, $x^2 = \psi$ the metric of the V_2 is given by $ds^2 = d\theta^2 + \sin^2 \theta d\psi^2$ and a brief calculation shows that the only non-vanishing Christoffel symbols of the second kind are $\left\{ {1 \atop 22} \right\} = - \sin \theta \cos \theta$ and $\left\{ {2 \atop 12} \right\} = \cot \theta$. Let us select the small circle $\theta = \alpha$ for the purpose of parallel propagation of the vector A^i. Along this circle $d\theta/dt = 0$, and consequently the equations (29.2) reduce to

$$\frac{dA^1}{d\psi} - \cos \alpha \sin \alpha\, A^2 = 0; \quad \frac{dA^2}{d\psi} + \cot \alpha\, A^1 = 0$$

whose solution can readily be obtained in the form

$$A^1 = \sin \alpha\, [c \sin (\psi \cos \alpha) + d \cos (\psi \cos \alpha)];$$
$$A^2 = c \cos (\psi \cos \alpha) - d \sin (\psi \cos \alpha),$$

where c and d are constants. Suppose we choose the vector A^i to be (1, 0) at the point defined by $\psi = 0$. Then by substitution we have $c = 0$ and $d = \operatorname{cosec} \alpha$. Thus the vector A^i is uniquely determined by the components $(\cos [\psi \cos \alpha], - \sin [\psi \cos \alpha]/\sin \alpha)$. The result of parallel propagation around the small circle then yields the vector $(\cos [2\pi \cos \alpha], - \sin [2\pi \cos \alpha]/\sin \alpha)$ which differs from the original vector (1, 0). The case of the great circle $\alpha = \pi/2$ is exceptional, as along it parallel propagation does lead back to the original vector.

Ex. In the V_2 with metric $ds^2 = du^2 + 2\lambda du dv + dv^2$, where λ is a function of u and v, show that the tangent vectors to the curves $u =$ constant form a field of parallel vectors along the curves $v =$ constant.

§ 30. Covariant derivative

By means of the concept of parallelism we are now able to supply a proof that the right-hand side of equation (23.3) constitutes a tensor. Consider a curve C determined by the equations $x^i = x^i(t)$. Choose p *arbitrary* contravariant vector-fields $X^i_{(1)}, X^i_{(2)}, \ldots X^i_{(p)}$ each parallel along the curve C and s *arbitrary* covariant vector-fields $Y_{(1)i}, Y_{(2)i}, \ldots Y_{(s)i}$ which are also parallel along the curve C. Then

$$\frac{dX^i_{(\beta)}}{dt} + \begin{Bmatrix} i \\ jk \end{Bmatrix} X^j_{(\beta)} \frac{dx^k}{dt} = 0 \quad (\beta = 1, 2, \ldots p), \tag{30.1}$$

and

$$\frac{dY_{(\alpha)i}}{dt} - \begin{Bmatrix} j \\ ik \end{Bmatrix} Y_{(\alpha)j} \frac{dx^k}{dt} = 0 \quad (\alpha = 1, 2, \ldots s). \tag{30.2}$$

Now let us set up the invariant

$$I \equiv A^{u_1 \ldots u_s}_{r_1 \ldots r_p} X^{r_1}_{(1)} X^{r_2}_{(2)} \ldots X^{r_p}_{(p)} Y_{(1)u_1} Y_{(2)u_2} \ldots Y_{(s)u_s}.$$

The derivative of this invariant with respect to t is also an invariant, and by application of (30.1) and (30.2) and appropriate changes of dummy indices we find

$$\frac{dI}{dt} = X^{r_1}_{(1)} \ldots X^{r_p}_{(p)} Y_{(1)u_1} \ldots Y_{(s)u_s} \left\{ \begin{aligned} & \frac{dA^{u_1 \ldots u_s}_{r_1 \ldots r_p}}{dt} \\ & + \sum_{\alpha=1}^{s} A^{u_1 \ldots u_{\alpha-1} k u_{\alpha+1} \ldots u_s}_{r_1 \ldots r_p} \begin{Bmatrix} u_\alpha \\ kn \end{Bmatrix} \frac{dx^n}{dt} \\ & - \sum_{\beta=1}^{p} A^{u_1 \ldots u_s}_{r_1 \ldots r_{\beta-1} l r_{\beta+1} \ldots r_p} \begin{Bmatrix} l \\ r_\beta n \end{Bmatrix} \frac{dx^n}{dt} \end{aligned} \right\}.$$

We deduce from the quotient law that the expression in brackets on the right-hand side of this equation is a tensor, which we call the intrinsic derivative, and denote by $\delta A^{u_1 \ldots u_s}_{r_1 \ldots r_p}/\delta t$. It follows immediately that

$$\frac{\delta A^{u_1 \ldots u_s}_{r_1 \ldots r_p}}{\delta t} = \frac{dx^n}{dt} \left[\frac{\partial A^{u_1 \ldots u_s}_{r_1 \ldots r_p}}{\partial x^n} + \sum_{\alpha=1}^{s} A^{u_1 \ldots u_{\alpha-1} k u_{\alpha+1} \ldots u_s}_{r_1 \ldots r_p} \begin{Bmatrix} u_\alpha \\ kn \end{Bmatrix} \right.$$
$$\left. - \sum_{\beta=1}^{p} A^{u_1 \ldots u_s}_{r_1 \ldots r_{\beta-1} l r_{\beta+1} \ldots r_p} \begin{Bmatrix} l \\ r_\beta n \end{Bmatrix} \right].$$

The quotient law again shows us that the expression in square brackets is a tensor, which we denote by $A^{u_1 \ldots u_s}_{r_1 \ldots r_p, n}$ and which we call the covariant derivative. This completes the required proof.

CHAPTER V

CURVATURE TENSOR

§ 31. Riemann-Christoffel tensor

We will now investigate the commutative problem with respect to covariant differentiation. Let us begin with the covariant derivative of an arbitrary covariant vector A_j

$$A_{j,n} = \frac{\partial A_j}{\partial x^n} - \begin{Bmatrix} l \\ jn \end{Bmatrix} A_l.$$

A further covariant differentiation yields

$$\begin{aligned} A_{j,np} &= \frac{\partial}{\partial x^p}(A_{j,n}) - \begin{Bmatrix} l \\ jp \end{Bmatrix} A_{l,n} - \begin{Bmatrix} l \\ np \end{Bmatrix} A_{j,l} \\ &= \frac{\partial^2 A_j}{\partial x^n \partial x^p} - \begin{Bmatrix} l \\ jn \end{Bmatrix} \frac{\partial A_l}{\partial x^p} - A_l \frac{\partial}{\partial x^p} \begin{Bmatrix} l \\ jn \end{Bmatrix} - \begin{Bmatrix} l \\ jp \end{Bmatrix} \frac{\partial A_l}{\partial x^n} \\ &+ \begin{Bmatrix} l \\ jp \end{Bmatrix} \begin{Bmatrix} k \\ ln \end{Bmatrix} A_k - \begin{Bmatrix} l \\ np \end{Bmatrix} \frac{\partial A_j}{\partial x^l} + \begin{Bmatrix} l \\ np \end{Bmatrix} \begin{Bmatrix} k \\ jl \end{Bmatrix} A_k. \end{aligned}$$

We interchange the indices n and p and subtract. After changing several dummy indices we have

$$A_{j,np} - A_{j,pn} =$$
$$\left[\frac{\partial}{\partial x^n} \begin{Bmatrix} l \\ jp \end{Bmatrix} - \frac{\partial}{\partial x^p} \begin{Bmatrix} l \\ jn \end{Bmatrix} + \begin{Bmatrix} l \\ ns \end{Bmatrix} \begin{Bmatrix} s \\ jp \end{Bmatrix} - \begin{Bmatrix} l \\ ps \end{Bmatrix} \begin{Bmatrix} s \\ jn \end{Bmatrix} \right] A_l.$$

Since A_l is an *arbitrary* vector, it follows from the quotient law that the expression in square brackets is a mixed tensor of the fourth order, of contravariant order one and covariant order three. Using the notation

$$(31.1)\quad R^{l}_{.jnp} \equiv \frac{\partial}{\partial x^n}\begin{Bmatrix} l \\ jp \end{Bmatrix} - \frac{\partial}{\partial x^p}\begin{Bmatrix} l \\ jn \end{Bmatrix} + \begin{Bmatrix} l \\ ns \end{Bmatrix}\begin{Bmatrix} s \\ jp \end{Bmatrix} - \begin{Bmatrix} l \\ ps \end{Bmatrix}\begin{Bmatrix} s \\ jn \end{Bmatrix},$$

we observe that $R^{l}_{.jnp}$ is a tensor of the fourth order, called the **Riemann-Christoffel** tensor. It is formed exclusively from the fundamental tensor g_{ij} and its derivatives up to and including the second order. This tensor does *not* depend on the choice of the vector A_i. We can now write

$$(31.2)\qquad A_{j,np} - A_{j,pn} = R^{l}_{.jnp} A_l .$$

It is clear from this equation, that the necessary and sufficient conditions that the covariant differentiation of *all* vectors be commutative, is that the Riemann-Christoffel tensor be identically zero. We stated in section 29 that the equations $A_{i,j} = 0$ are not as a rule consistent. In fact, equations (31.2) show us that a necessary, but not sufficient condition, for the consistency of $A_{i,j} = 0$ is that $R^{l}_{.jnp} A_l = 0$, a set of equations which is not generally satisfied.

Referring to the definition (31.1), we observe that

$$(31.3)\qquad R^{l}_{.jnp} = - R^{l}_{.jpn}.$$

That is, $R^{l}_{.jnp}$ is skew-symmetric with respect to the indices n and p.

Ex. 1. Prove that $R^{l}_{.jnp} + R^{l}_{.npj} + R^{l}_{.pjn} = 0$.

Ex. 2. Prove that $R^{l}_{.lnp} = 0$.

§ 32. Curvature tensor

We now introduce the covariant **curvature tensor** defined by

$$(32.1)\qquad R_{rjnp} = g_{rl} R^{l}_{.jnp} .$$

On substituting from (31.1) into (32.1) we obtain

$$R_{rjnp} = \frac{\partial}{\partial x^n}\left[g_{rl}\begin{Bmatrix} l \\ jp \end{Bmatrix}\right] - \frac{\partial g_{rl}}{\partial x^n}\begin{Bmatrix} l \\ jp \end{Bmatrix} - \frac{\partial}{\partial x^p}\left[g_{rl}\begin{Bmatrix} l \\ jn \end{Bmatrix}\right]$$
$$+ \frac{\partial g_{rl}}{\partial x^p}\begin{Bmatrix} l \\ jn \end{Bmatrix} + g_{rl}\begin{Bmatrix} l \\ ns \end{Bmatrix}\begin{Bmatrix} s \\ jp \end{Bmatrix} - g_{rl}\begin{Bmatrix} l \\ ps \end{Bmatrix}\begin{Bmatrix} s \\ jn \end{Bmatrix},$$

which reduces on application of (20.3) and (20.4) to

$$R_{rjnp} = \frac{\partial}{\partial x^n}[jp, r] - \frac{\partial}{\partial x^p}[jn, r] + \begin{Bmatrix} l \\ jn \end{Bmatrix}[rp, l] - \begin{Bmatrix} l \\ jp \end{Bmatrix}[rn, l].$$

We can reduce this further by means of (20.1) and (20.2) and finally obtain the important formula

$$(32.2) \qquad R_{rjnp} = \frac{1}{2}\left(\frac{\partial^2 g_{rp}}{\partial x^j \partial x^n} + \frac{\partial^2 g_{jn}}{\partial x^r \partial x^p} - \frac{\partial^2 g_{rn}}{\partial x^j \partial x^p} - \frac{\partial^2 g_{jp}}{\partial x^r \partial x^n}\right)$$
$$+ g^{ts}([jn, s][rp, t] - [jp, s][rn, t]).$$

From this result we immediately deduce the relations

$$(32.3) \qquad \begin{cases} R_{rjnp} = -R_{jrnp}, \\ R_{rjnp} = -R_{rjpn}, \\ R_{rjnp} = R_{nprj}, \end{cases}$$

and

$$(32.4) \qquad R_{rjnp} + R_{rnpj} + R_{rpjn} = 0.$$

We are now faced by this problem: how many distinct arithmetical non-vanishing components does the tensor R_{rjnp} generally possess? Referring to (32.3) we see that a component is zero if either $r = j$ or $n = p$. Thus the components apart from sign conform to the three types R_{rjrj}, R_{rjrp} and R_{rjnp} where r, j, n and p are distinct from one another. There are as many components of the type R_{rjrj} as there are ways of combining r and j. That is, $\frac{1}{2}N(N-1)$. There are as many components of the type R_{rjrp} as there are ways of combining j and p after selection of r. That is, $\frac{1}{2}N(N-1)(N-2)$. The number of combinations of r, j, n and p is $\frac{1}{24}N(N-1)(N-2)(N-3)$. But R_{rjnp} is determined, except for sign, when r is paired with either j, n or p. Thus there are $\frac{1}{8}N(N-1)(N-2)(N-3)$ components of the type R_{rjnp}. Equations (32.3) enable us to rewrite (32.4) in the form

$$R_{jrnp} + R_{jprn} + R_{jnpr} = 0.$$

And so with a given set of four indices, only one independent equation of the type (32.4) exists. Further, if the indices j, r, n and p are not distinct, (32.4) reduces to one of the equations (32.3). Hence the number of independent equations of the type (32.4) is $\frac{1}{24}N(N-1)(N-2)(N-3)$. Therefore the number of distinct components of R_{rjnp} is

$$\tfrac{1}{2}N(N-1)+\tfrac{1}{2}N(N-1)(N-2)+\tfrac{1}{8}N(N-1)(N-2)(N-3)$$
$$-\tfrac{1}{24}N(N-1)(N-2)(N-3) = \tfrac{1}{12}N^2(N^2-1).$$

In particular we see that the curvature tensor of a V_2 has only one distinct non-vanishing component.

Ex. 1. Prove the relation (32.4) by setting up a geodesic coordinate system.

Ex. 2. Prove that $R_{1212} = -G\dfrac{\partial^2 G}{\partial u^2}$ for the V_2 whose line-element is $ds^2 = du^2 + G^2dv^2$, where G is a function of u and v.

§ 33. Ricci tensor - Curvature invariant

At first sight there appear to be three different ways of contracting the Riemann-Christoffel tensor $R^l_{.jnp}$. We have $R^l_{.lnp} = g^{ls}R_{slnp} = 0$, because R_{slnp} is skew-symmetric in s and l. We see from (31.3) that $R^l_{.jnl} = -R^l_{.jln}$. Hence we need only consider the contraction, called the **Ricci tensor**, defined by

$$R_{jn} = R^l_{.jnl} = g^{ls}R_{sjnl}. \tag{33.1}$$

On contraction of l and p in (31.1) and on substitution from (20.6) we find that

$$R_{jn} = \frac{\partial^2}{\partial x^j \partial x^n}\{\log \sqrt{g}\} - \frac{\partial}{\partial x^l}\begin{Bmatrix} l \\ jn \end{Bmatrix} \tag{33.2}$$
$$+ \begin{Bmatrix} l \\ ns \end{Bmatrix}\begin{Bmatrix} s \\ jl \end{Bmatrix} - \begin{Bmatrix} s \\ jn \end{Bmatrix}\frac{\partial}{\partial x^s}\{\log \sqrt{g}\},$$

from which it is clear that R_{jn} is symmetric. (If g is negative, we must replace $\log \sqrt{g}$ by $\log \sqrt{-g}$.)

The **curvature invariant** is defined by

$$R = g^{jn} R_{jn}. \tag{33.3}$$

A space for which $R_{ij} = I g_{ij}$ at all points, where I is an invariant, is called an **Einstein** space. Inner multiplication by g^{ij} shows that $R = NI$. Thus for an Einstein space

$$R_{ij} = \frac{1}{N} R g_{ij}. \tag{33.4}$$

Ex. For a V_2, prove that $gR_{ij} = -g_{ij}R_{1212}$ and $gR = -2R_{1212}$. Hence deduce that every V_2 is an Einstein space.

§ 34. Bianchi's identity

Let us choose a system of geodesic coordinates. On referring to (31.1) we have by covariant differentiation that

$$R^l_{.jnp,r} = \frac{\partial}{\partial x^r}(R^l_{.jnp}) = \frac{\partial^2}{\partial x^r \partial x^n}\begin{Bmatrix} l \\ jp \end{Bmatrix} - \frac{\partial^2}{\partial x^p \partial x^r}\begin{Bmatrix} l \\ jn \end{Bmatrix}$$

at the pole. Cyclic interchange of n, p and r gives us two other equations. We obtain by addition

$$R^l_{.jnp,r} + R^l_{.jpr,n} + R^l_{.jrn,p} = 0. \tag{34.1}$$

This is a tensor equation, true at the pole of a geodesic system of coordinates. Thus it also holds for every coordinate system at that pole. Further any point can be chosen as the pole of a geodesic coordinate system. Therefore equation (34.1) is true at all points of space. Inner multiplication by g_{lm} yields the Bianchi identity

$$R_{mjnp,r} + R_{mjpr,n} + R_{mjrn,p} = 0. \tag{34.2}$$

The Einstein tensor is defined by

$$G^i_{.j} = g^{il} R_{jl} - \tfrac{1}{2} R \delta^i_j. \tag{34.3}$$

The inner multiplication of (34.2) by $g^{mp}g^{jn}$ and the application of (33.1), (33.3), and (32.3) give us the equation

$$R_{,r} - g^{jn} R_{jr,n} - g^{mp} R_{mr,p} = 0,$$

which may be written

$$R_{,r} = 2g^{jn} R_{jr,n}. \tag{34.4}$$

Differentiating (34.3) covariantly, we obtain

$$G^i_{.j,i} = g^{il} R_{jl,i} - \tfrac{1}{2} R_{,i} \delta^i_j = g^{il} R_{jl,i} - \tfrac{1}{2} R_{,j}.$$

That is

$$G^i_{.j,i} = 0. \tag{34.5}$$

This equation is important in the theory of Relativity.

§ 35. Riemannian curvature

From any two vectors A^i and B^i at a point of a V_N, we can construct the invariant $R_{rjnp}A^rA^nB^jB^p$. Let us consider what happens if we replace the vectors A^i and B^i by the two linear combinations

$$X^i = \lambda A^i + \mu B^i, \quad Y^i = \rho A^i + \tau B^i,$$

where λ, μ, ρ and τ are invariants. A straightforward calculation, with the aid of (32.3), shows that

$$R_{rjnp} X^r X^n Y^j Y^p = (\lambda\tau - \rho\mu)^2 R_{rjnp} A^r A^n B^j B^p.$$

Thus the expression $R_{rjnp}A^rA^nB^jB^p$, which is an invariant with respect to coordinate transformations, is *almost* an invariant under linear transformations of vectors. In order to obtain an expression which is also invariant under linear transformations of vectors let us evaluate

$$
\begin{aligned}
&(g_{rn}g_{jp}-g_{rp}g_{jn})X^rX^nY^jY^p \\
&= (\lambda A_n+\mu B_n)(\lambda A^n+\mu B^n)(\rho A_p+\tau B_p)(\rho A^p+\tau B^p) \\
&\qquad -(\lambda A_p+\mu B_p)(\rho A^p+\tau B^p)(\lambda A_j+\mu B_j)(\rho A^j+\tau B^j) \\
&= (e_A\lambda^2A^2+e_B\mu^2B^2+2\lambda\mu\cos\theta\, AB)(e_A\rho^2A^2+e_B\tau^2B^2+2\rho\tau\cos\theta\, AB) \\
&\qquad -(e_A\lambda\rho A^2+e_B\mu\tau B^2+[\lambda\tau+\rho\mu]\cos\theta\, AB)^2 \\
&= (\lambda\tau-\mu\rho)^2(e_Ae_B-\cos^2\theta)A^2B^2 \\
&= (\lambda\tau-\mu\rho)^2(g_{rn}g_{jp}-g_{rp}g_{jn})A^rA^nB^jB^p,
\end{aligned}
$$

where θ is the angle between the vectors A^i and B^i. It follows that

$$
K = \frac{R_{rjnp}A^rA^nB^jB^p}{(g_{rn}g_{jp}-g_{rp}g_{jn})A^rA^nB^jB^p} \tag{35.1}
$$

is an invariant which is unaltered at a point, when the two vectors determining it are replaced by any linear combination. This invariant is called the **Riemannian curvature** of the space V_N associated with the vectors A^i and B^i. Note that the denominator of K is unity if the vectors A^i and B^i are orthogonal unit vectors.

At any point of a two-dimensional space there exist only two independent vectors. Hence the Riemannian curvature of a V_2 is uniquely determined at each point. Its value is easily found by choosing the two vectors whose components are (1, 0) and (0, 1) respectively. Then

$$
K = \frac{R_{1212}}{g_{11}g_{22}-g_{12}{}^2} = \frac{R_{1212}}{g}. \tag{35.2}
$$

§ 36. Flat space

We say that a space is **flat** if $K = 0$ at every point of it. From (35.1) the necessary and sufficient condition is

$$
R_{rjnp}A^rA^nB^jB^p = 0
$$

for *all* vectors A^i and B^i. In virtue of equations (32.3) it follows that

$$
R_{rjnp} + R_{njrp} + R_{nprj} + R_{rpnj} = 0.
$$

That is,

$$R_{rjnp} + R_{rpnj} = 0,$$

which can be written

$$R_{rjnp} = R_{rpjn}.$$

Interchanging j, n and p cyclically we obtain

$$R_{rnpj} = R_{rjnp}.$$

Thus we have

$$R_{rjnp} = R_{rpjn} = R_{rnpj}.$$

Substitution in (32.4) immediately yields

$$R_{rjnp} = 0.$$

Conversely, if $R_{rnjp} = 0$, then it is clear that $K = 0$. Therefore, the necessary and sufficient condition that a space V_N be flat is that the Riemann-Christoffel tensor be identically zero.

In a flat space $R^l_{.jnp} = 0$, hence we deduce from (31.2) that in a flat space the equations $A_{i,j} = 0$ are consistent. Inner multiplication by dx^j/dt yields $\delta A_i/\delta t = 0$. Thus in a flat space the property of parallelism is independent of the choice of a curve. We may therefore say that parallelism is an absolute property of a flat space.

A familiar example of a flat space is the Euclidean plane for which the metric is $ds^2 = dx^2 + dy^2$ in rectangular cartesian and $ds^2 = dr^2 + r^2 d\theta^2$ in polar coordinates.

Ex. If the metric of a two-dimensional flat space is $f(r)[(dx^1)^2 + (dx^2)^2]$, where $(r)^2 = (x^1)^2 + (x^2)^2$, show that $f(r) = c(r)^k$, where c and k are constants.

§ 37. Space of constant curvature

Let us now investigate spaces in which the Riemannian curvature at every point does not depend on the choice of the associated vectors A^i and B^i. From (35.1) the necessary and sufficient condition is that

$$\{K(g_{rn}g_{jp} - g_{rp}g_{jn}) - R_{rjnp}\}A^r A^n B^j B^p = 0$$

for *all* vectors A^i and B^i. A similar calculation to that of the previous section will show that this condition reduces to

$$R_{rjnp} = K(g_{rn}g_{jp} - g_{rp}g_{jn}),$$

where K is now a function of the coordinates x^i.

Covariant differentiation gives us

$$R_{rjnp,t} = K_{,t}(g_{rn}g_{jp} - g_{rp}g_{jn}).$$

We substitute this result in Bianchi's identity (34.2) and obtain

$$K_{,r}(g_{mn}g_{jp} - g_{mp}g_{jn}) + K_{,n}(g_{mp}g_{jr} - g_{mr}g_{jp}) + K_{,p}(g_{mr}g_{jn} - g_{mn}g_{jr}) = 0.$$

Inner multiplication by $g^{mn}g^{jp}$ yields

$$(N-1)(N-2)K_{,r} = 0.$$

Hence if $N > 2$, it follows that K is constant. Thus we have proved Schur's theorem; 'if at each point of a space V_N, $(N > 2)$, the Riemannian curvature is a function of the coordinates only, then it is constant throughout the V_N'. Such a V_N is called a **space of constant curvature.**

The metric of the V_2 formed by the surface of a sphere of radius a is $ds^2 = a^2(d\theta^2 + \sin^2\theta d\psi^2)$ in spherical polar coordinates. The reader is asked to verify that $R_{1212} = a^2 \sin^2\theta$ and that it follows from (35.2) that the surface of a sphere is a surface of constant curvature $1/a^2$.

Ex. 1. Show that a space of constant curvature is an Einstein space.

Ex. 2. In a Euclidean V_4, prove that the hypersphere

$$x^1 = c\sin\theta\sin\varphi\sin\psi, \quad x^2 = c\sin\theta\sin\varphi\cos\psi$$
$$x^3 = c\sin\theta\cos\varphi, \quad x^4 = c\cos\theta$$

is a V_3 of constant curvature $1/c^2$.

CHAPTER VI

EUCLIDEAN THREE-DIMENSIONAL DIFFERENTIAL GEOMETRY

Euclidean Geometry investigates the properties of figures which are invariant with respect to translations and rotations in space. It may be subdivided into Algebraic and Differential Geometry. The former studies by algebraic methods the theory applicable to entire configurations such as the class or degree of a curve. The latter discusses by means of the Calculus those properties which depend on a restricted portion of the figure. For example, the total curvature of a surface at a point only depends on the shape of the surface at that point. Succinctly we may say that Differential Geometry is the study of geometry *in the small.* This chapter is not intended to be a complete course on the subject. However, sufficient theory is developed to indicate the scope and power of the tensor method.

§ 38. Permutation tensors

In the Euclidean three-dimensional space let us introduce the quantities defined by

$$(38.1) \qquad \varepsilon_{ijk} = \sqrt{g}\, e_{ijk}; \quad \varepsilon^{ijk} = \frac{1}{\sqrt{g}}\, e_{ijk},$$

where e_{ijk} are the **permutation symbols** defined by the following conditions:-

(i) $e_{ijk} = 0$ if any two of the indices i, j and k are equal,
(ii) $e_{123} = e_{231} = e_{312} = +1,$
(iii) $e_{132} = e_{321} = e_{213} = -1,$

and g is the determinant formed from the fundamental tensor g_{ij} of the space referred to some general coordinate system, which is not necessarily rectangular cartesian. The definitions show us that e_{ijk}, ε_{ijk} and ε^{ijk} are skew-symmetric in all their indices. (Note that in this chapter the range of Latin indices is now from 1 to 3).

We shall first prove that although e_{ijk} is not a tensor, both ε_{ijk} and ε^{ijk} are tensors. We observe that

$$e_{ijk}\frac{\partial x^i}{\partial \bar{x}^l}\frac{\partial x^j}{\partial \bar{x}^m}\frac{\partial x^k}{\partial \bar{x}^n} = e_{jik}\frac{\partial x^j}{\partial \bar{x}^l}\frac{\partial x^i}{\partial \bar{x}^m}\frac{\partial x^k}{\partial \bar{x}^n} = -e_{ijk}\frac{\partial x^i}{\partial \bar{x}^m}\frac{\partial x^j}{\partial \bar{x}^l}\frac{\partial x^k}{\partial \bar{x}^n}.$$

Thus $e_{ijk}\dfrac{\partial x^i}{\partial \bar{x}^l}\dfrac{\partial x^j}{\partial \bar{x}^m}\dfrac{\partial x^k}{\partial \bar{x}^n}$ is skew-symmetric in l and m. Similarly it is skew-symmetric in the indices l, m and n. But this expression apart from sign is the Jacobian determinant $\left|\dfrac{\partial x^r}{\partial \bar{x}^s}\right|$. It therefore follows from the theory of determinants that

$$e_{ijk}\frac{\partial x^i}{\partial \bar{x}^l}\frac{\partial x^j}{\partial \bar{x}^m}\frac{\partial x^k}{\partial \bar{x}^n} = e_{lmn}\left|\frac{\partial x^r}{\partial \bar{x}^s}\right|.$$

Now the fundamental tensor g_{ij} transforms to $\bar{g}_{ij}$ when we transform to the coordinate system $\bar{x}^i$. We find from (21.1) that their determinants satisfy the equation

$$\bar{g} = g\left|\frac{\partial x^r}{\partial \bar{x}^s}\right|^2.$$

The quantities ε_{ijk} transform to $\bar{\varepsilon}_{ijk}$ where

$$\bar{\varepsilon}_{lmn} = \sqrt{\bar{g}}\, e_{lmn} = \sqrt{g}\, e_{lmn}\left|\frac{\partial x^r}{\partial \bar{x}^s}\right| = \sqrt{g}\, e_{ijk}\frac{\partial x^i}{\partial \bar{x}^l}\frac{\partial x^j}{\partial \bar{x}^m}\frac{\partial x^k}{\partial \bar{x}^n}.$$

Hence

$$\bar{\varepsilon}_{lmn} = \varepsilon_{ijk}\frac{\partial x^i}{\partial \bar{x}^l}\frac{\partial x^j}{\partial \bar{x}^m}\frac{\partial x^k}{\partial \bar{x}^n}, \tag{38.2}$$

from which we see that ε_{ijk} is a covariant tensor of the third order. Also

$$\varepsilon^{lmn} = \frac{1}{\sqrt{g}} e_{lmn} = \frac{1}{\sqrt{\bar{g}}} \left| \frac{\partial x^r}{\partial \bar{x}^s} \right| e_{lmn} = \frac{1}{\sqrt{\bar{g}}} e_{ijk} \frac{\partial x^l}{\partial \bar{x}^i} \frac{\partial x^m}{\partial \bar{x}^j} \frac{\partial x^n}{\partial \bar{x}^k};$$

therefore,

$$\varepsilon^{lmn} = \bar{\varepsilon}^{ijk} \frac{\partial x^l}{\partial \bar{x}^i} \frac{\partial x^m}{\partial \bar{x}^j} \frac{\partial x^n}{\partial \bar{x}^k},$$

which shows that ε^{lmn} is a contravariant tensor of the third order. We call ε_{ijk} and ε^{ijk} the **permutation tensors.**

If the coordinate system is rectangular cartesian, $g = 1$ and the permutation tensors have as their components the permutation symbols. In this coordinate system the covariant derivatives $\varepsilon_{ijk,l}$ and $\varepsilon^{ijk}{}_{,l}$, which are both tensors, are zero. Thus the covariant derivatives $\varepsilon_{ijk,l}$ and $\varepsilon^{ijk}{}_{,l}$ are zero tensors in all coordinate systems. Hence the permutation tensors behave like constants with respect to covariant differentiation.

From a covariant vector A_i we can form the vector

$$B^j = \varepsilon^{jkl} A_{l,k}$$

and we call B^j the **curl** of the vector A_i and write it curl A_i.

Ex. 1. Prove directly that $\varepsilon_{ijk,l} = 0$.

Ex. 2. Prove that $\varepsilon_{ijk} = g_{il} g_{jm} g_{kn} \varepsilon^{lmn}$.

Ex. 3. Show that the components of curl A_i are

$$\frac{1}{\sqrt{g}} \left(\frac{\partial A_3}{\partial x^2} - \frac{\partial A_2}{\partial x^3} \right), \quad \frac{1}{\sqrt{g}} \left(\frac{\partial A_1}{\partial x^3} - \frac{\partial A_3}{\partial x^1} \right), \quad \frac{1}{\sqrt{g}} \left(\frac{\partial A_2}{\partial x^1} - \frac{\partial A_1}{\partial x^2} \right).$$

§ 39. Vector product

We can form the contravariant vector

$$C^i = \varepsilon^{ijk} A_j B_k \tag{39.1}$$

from the two covariant vectors A_i and B_i. To find the geometrical interpretation of the vector C^i, let us choose a rectangular cartesian coordinate system. In such a system the distinction between contravariance and covariance disappears and the C^i are now the components of the vector product* of A_i and B_i. Thus C^i is a vector of magnitude $AB \sin \theta$, orthogonal to both vectors A_i and B_i, where θ is the angle between these vectors. Its direction is uniquely determined by the fact that A_i, B_i and C^i form a right-handed system.

§ 40. Frenet formulae

In this section we shall investigate the theory of twisted curves. Let the curve be given by the equations $x^i = x^i(s)$, where the parameter s measures the arc-distance along the curve. Then the unit tangent vector T^i to the curve is $T^i \equiv dx^i/ds$ and satisfies the equation $g_{ij}T^iT^j = 1$. On differentiation we obtain $g_{ij}\dfrac{\delta T^i}{\delta s}T^j = 0$. That is, $\delta T^i/\delta s$ is a vector orthogonal to the tangent vector. Its magnitude κ is called the **curvature** of the curve, and the unit vector

$$N^i = \frac{1}{\kappa}\frac{\delta T^i}{\delta s} \tag{40.1}$$

is called the unit **principal normal** vector. We define the unit **binormal** vector B^i to be the unit vector orthogonal to both the tangent and principal normal vectors, and oriented so that the tangent, principal normal and binormal vectors form a right-handed system. Hence from (39.1) we have

$$B^i = \varepsilon^{ijk}T_jN_k. \tag{40.2}$$

Since the vectors T^i and N^i are orthogonal, $g_{ij}T^iN^j = 0$. We differentiate intrinsically and obtain

$$g_{ij}T^i\frac{\delta N^j}{\delta s} + g_{ij}\frac{\delta T^i}{\delta s}N^j = 0.$$

* D. E. Rutherford, Vector Methods, p. 7.

We substitute for $\frac{\delta T^i}{\delta s}$ from (40.1) and then replace $g_{ij}N^iN^j$ by $g_{ij}T^iT^j$, both being equal to unity, and obtain

$$g_{ij}T^i\left(\frac{\delta N^j}{\delta s} + \kappa T^j\right) = 0.$$

Now we differentiate $g_{ij}N^iN^j = 1$, which gives us $g_{ij}N^i\frac{\delta N^j}{\delta s} = 0$. Thus

$$g_{ij}N^i\left(\frac{\delta N^j}{\delta s} + \kappa T^j\right) = 0.$$

Therefore the vector $\frac{\delta N^j}{\delta s} + \kappa T^j$ is orthogonal to both the tangent vector T^i and the principal normal vector N^i. Accordingly it is in the direction of the binormal vector B^j, and we can write

$$B^j = \frac{1}{\tau}\left(\frac{\delta N^j}{\delta s} + \kappa T^j\right). \tag{40.3}$$

The invariant τ thus introduced* is called the **torsion** of the curve. Note that it may be positive or negative in contrast to the curvature which is essentially positive, being the magnitude of a vector.

We differentiate (40.2) intrinsically and substitute from (40.1) and (40.3) and the result is

$$\begin{aligned}\frac{\delta B^i}{\delta s} &= \varepsilon^{ijk}\frac{\delta T_j}{\delta s}N_k + \varepsilon^{ijk}T_j\frac{\delta N_k}{\delta s}\\ &= \kappa\varepsilon^{ijk}N_jN_k + \varepsilon^{ijk}T_j[\tau B_k - \kappa T_k]\end{aligned}$$

which reduces on account of the skew-symmetry of ε^{ijk} to

$$\frac{\delta B^i}{\delta s} = \tau\varepsilon^{ijk}T_jB_k.$$

But T_i, N_i and B_i form a right-handed system of unit

* Some text books use τ to denote the reciprocal of the torsion.

vectors; and so $\varepsilon^{ijk} T_j B_k = -N^i$ and our equations become

$$\frac{\delta B^i}{\delta s} = -\tau N^i. \tag{40.4}$$

Equations (40.1), (40.3) and (40.4) are called the **Frenet** formulae. On account of their importance in the theory of curves we group these formulae together for convenient reference in the form

$$\begin{cases} \dfrac{\delta T^i}{\delta s} = \qquad \kappa N^i \\ \dfrac{\delta N^i}{\delta s} = -\kappa T^i \qquad + \tau B^i \\ \dfrac{\delta B^i}{\delta s} = \qquad -\tau N^i. \end{cases} \tag{40.5}$$

If the coordinate system is rectangular cartesian, the intrinsic derivatives become the ordinary derivatives and we have the well-known Frenet formulae.

Ex. 1. Show that $\kappa^2 = g_{ij} \dfrac{\delta T^i}{\delta s} \dfrac{\delta T^j}{\delta s}$.

Ex. 2. Prove that $\tau = \varepsilon^{ijk} T_i N_j \dfrac{\delta N_k}{\delta s}$.

Ex. 3. A **helix** is defined to be a curve whose tangent vector makes a constant angle with a fixed direction. Prove that the necessary and sufficient condition that a curve be a helix is that the ratio of the torsion to the curvature is constant.

Ex. 4. Prove that $\varepsilon^{ijk} \dfrac{\delta T_i}{\delta s} \dfrac{\delta^2 T_j}{\delta s^2} \dfrac{\delta^3 T_k}{\delta s^3} = \kappa^5 \dfrac{d}{ds}\left(\dfrac{\tau}{\kappa}\right)$. Hence a curve is a helix if and only if $\varepsilon^{ijk} \dfrac{\delta T_i}{\delta s} \dfrac{\delta^2 T_j}{\delta s^2} \dfrac{\delta^3 T_k}{\delta s^3} = 0$.

§ 41. Surface - First fundamental form

We will now investigate the theory of surfaces. The three equations

$$x^1 = x^1(u^1, u^2); \quad x^2 = x^2(u^1, u^2); \quad x^3 = x^3(u^1, u^2), \tag{41.1}$$

where u^1 and u^2 are parameters and the x^i are three functions of u^1 and u^2 which are real and continuous, generally represent a surface. These equations may be briefly written $x^i = x^i(u^\alpha)$ if we adopt the convention that Greek indices will always have the range 1, 2. A point at which the Jacobian matrix $\left[\dfrac{\partial x^i}{\partial u^\alpha}\right]$ is of rank *two* is called a *regular* point. A point may fail to be regular either because it is a singularity on the surface, for example, the vertex of a cone, or because it is a singularity of the parametric representation, for example, the poles of a sphere. (See equation (41.2)).

Each pair of values of u^α determines a point on the surface. That is, the u^α form a coordinate system upon it. Thus an equation of the type $f(u^1, u^2) = 0$ must define a curve. Although a unique point on the surface corresponds to a fixed pair of values of u^α, the converse is not necessarily true. This is illustrated by the equations

$$(41.2)\quad x^1 = a \sin u^1 \cos u^2; \; x^2 = a \sin u^1 \sin u^2; \; x^3 = a \cos u^1,$$

which specify the surface of a sphere of radius a. In practice we restrict the range of the parameters so that the converse will also be valid. In our example, the parameters are limited to the ranges $0 \leqq u^1 \leqq \pi$ and $0 \leqq u^2 < 2\pi$. Then, except for the two poles, there corresponds to each point of the sphere a unique pair of values of u^α. The equations $u^1 = 0$ and $u^1 = \pi$ represent, not curves, but the poles respectively. At the poles, the coordinate u^2 is indeterminate.

We shall exclude singular points from our discussion by considering only that portion of a surface on which there is a unique correspondence between its points and pairs of values of the coordinates u^α. That is, in this portion every curve of the family u^1 = constant intersects every curve of the family u^2 = constant in one point only. The curves u^1 = constant are called **u^2-curves** and the curves $u^2 =$ constant the **u^1-curves.** Collectively they are designated the **coordinate** or **parametric curves** on the surface. Along a coordinate curve we choose the positive direction to correspond with increasing values of the variables u^1 or u^2 respectively.

The contravariant vectors dx^i and du^α which represent the same displacement in space and on the surface respectively are connected by the equations

$$dx^i = \frac{\partial x^i}{\partial u^\alpha} du^\alpha \tag{41.3}$$

where we extend the summation convention to apply to Greek indices. Hence the line-element ds on the surface is given by

$$ds^2 = g_{ij}\, dx^i\, dx^j = g_{ij} \frac{\partial x^i}{\partial u^\alpha} \frac{\partial x^j}{\partial u^\beta} du^\alpha\, du^\beta.$$

Let us write

$$a_{\alpha\beta} = g_{ij} \frac{\partial x^i}{\partial u^\alpha} \frac{\partial x^j}{\partial u^\beta}, \tag{41.4}$$

from which it is clear that $a_{\alpha\beta}$ is symmetric, and so we have

$$ds^2 = a_{\alpha\beta}\, du^\alpha\, du^\beta. \tag{41.5}$$

To this equation we apply the quotient law. Since $a_{\alpha\beta}$ is symmetric it follows that $a_{\alpha\beta}$ is a covariant tensor with respect to transformations of the coordinate system u^α. We call it the **fundamental surface tensor.** Also $a_{\alpha\beta}\, du^\alpha\, du^\beta$ is named the **first fundamental form** of the surface.

Let us select rectangular cartesian coordinates in space. Then we readily obtain the metric in the well-known form

$$ds^2 = E\, du^2 + 2F\, du dv + G\, dv^2,$$

where $u = u^1$, $v = u^2$ and

$$E = \sum_i \frac{\partial x^i}{\partial u} \frac{\partial x^i}{\partial u}, \quad F = \sum_i \frac{\partial x^i}{\partial u} \frac{\partial x^i}{\partial v}, \quad G = \sum_i \frac{\partial x^i}{\partial v} \frac{\partial x^i}{\partial v}.$$

§ 42. Surface vectors

When we transform coordinates in space, dx^i is a contravariant vector but du^α is an invariant. On the other hand, if we transform the coordinates u^α on the surface, dx^i is an invariant but du^α is a contravariant vector. Thus the equations (41.3) indicate that we can regard $\partial x^i/\partial u^\alpha$ as both a contravariant space vector and a covariant surface vector. We may therefore introduce the notation

$$x^i_\alpha \equiv \frac{\partial x^i}{\partial u^\alpha} \tag{42.1}$$

and rewrite (41.4) in the form

$$a_{\alpha\beta} = g_{ij} x^i_\alpha x^j_\beta. \tag{42.2}$$

A curve on the surface is represented parametrically by the equations $u^\alpha = u^\alpha(t)$. The vector du^α/dt is a tangent vector to this curve. Its space components are then given by the equations

$$\frac{dx^i}{dt} = \frac{\partial x^i}{\partial u^\alpha}\frac{du^\alpha}{dt} = x^i_\alpha \frac{du^\alpha}{dt}. \tag{42.3}$$

But if the components of the space tangent vector dx^i/dt are fixed, (42.3) consists of three equations for the two unknowns du^α/dt. They are not consistent unless the vector lies in the surface, when a unique solution would exist.

Next we consider a surface vector-field A^α. We can set up a unique curve C on the surface by the differential equations $du^\alpha/dt = A^\alpha$, provided that the vector-field is fixed at some particular point. Then A^α is the surface tangent vector to this curve C. Let us designate the space components of A^α by A^i, and so equations (42.3) state that these components are connected by the relations

$$A^i = x^i_\alpha A^\alpha. \tag{42.4}$$

The magnitude of the vector A^i is given by

$$(A)^2 = g_{ij}A^iA^j = g_{ij}x^i_\alpha x^j_\beta A^\alpha B^\beta.$$

That is,

$$(A)^2 = a_{\alpha\beta}A^\alpha A^\beta. \tag{42.5}$$

In particular, du^α/ds is the unit tangent vector to a curve on the surface if the parameter s measures arc-distance along it.

The angle θ between the two *unit* vectors A^i and B^i is obtained from

$$\cos\theta = g_{ij}A^iB^j = g_{ij}x^i_\alpha x^j_\beta A^\alpha B^\beta.$$

That is,

$$\cos\theta = a_{\alpha\beta}A^\alpha B^\beta. \tag{42.6}$$

It follows that the necessary and sufficient condition for the orthogonality of two surface vectors A^α and B^β is

$$a_{\alpha\beta}A^\alpha B^\beta = 0. \tag{42.7}$$

We see from equations (42.5), (42.6) and (42.7) that the familiar formulae apply equally well on the surface provided that we employ in them the components of the vector in the surface and the surface fundamental tensor. Also we can raise and lower indices of surface tensors in the usual way with the surface fundamental tensor $a_{\alpha\beta}$ and its conjugate symmetric tensor $a^{\alpha\beta}$. These two tensors are connected by the equations

$$a_{\alpha\beta}a^{\alpha\gamma} = \delta^\gamma_\beta, \tag{42.8}$$

where δ^γ_β is the two-dimensional Kronecker delta. It is worthy of note that

$$a^{11} = a_{22}/a;\quad a^{12} = a^{21} = -a_{12}/a;\quad a^{22} = a_{11}/a, \tag{42.9}$$

where

$$a = a_{11}a_{22} - (a_{12})^2.$$

Ex. Prove that $A_\alpha = x_\alpha^r A_r$.

§ 43. Permutation surface tensor

In section 38 we introduced the permutation tensors in space. Similarly we introduce on the surface the quantities defined by

$$\varepsilon_{\alpha\beta} = \sqrt{a}\, e_{\alpha\beta}; \quad \varepsilon^{\alpha\beta} = \frac{1}{\sqrt{a}}\, e_{\alpha\beta}, \tag{43.1}$$

where

$$e_{11} = e_{22} = 0; \quad e_{12} = +1; \quad e_{21} = -1.$$

It is left as an exercise for the reader to show that $\varepsilon_{\alpha\beta}$ and $\varepsilon^{\alpha\beta}$ are a covariant and a contravariant skew-symmetric surface tensor respectively. They are called the **surface permutation tensors.** We see that $\varepsilon_{\alpha\beta}$ can be obtained from $\varepsilon^{\alpha\beta}$ by lowering the indices since $\varepsilon_{\alpha\beta} = a_{\alpha\gamma}\, a_{\beta\delta}\, \varepsilon^{\gamma\delta}$.

Now we derive an important formula for the angle θ between two *unit* vectors A^α and B^α. This angle is given by $\cos\theta = a_{\alpha\beta} A^\alpha B^\beta$. Accordingly, using (42.6)

$$\begin{aligned}
\sin^2\theta &= 1 - a_{\alpha\beta} A^\alpha B^\beta a_{\gamma\delta} A^\gamma B^\delta \\
&= (a_{\alpha\gamma}\, a_{\beta\delta} - a_{\alpha\beta}\, a_{\gamma\delta}) A^\alpha A^\gamma B^\beta B^\delta \\
&= a e_{\alpha\delta}\, e_{\gamma\beta}\, A^\alpha A^\gamma B^\beta B^\delta \\
&= \varepsilon_{\alpha\delta}\, \varepsilon_{\gamma\beta}\, A^\alpha A^\gamma B^\beta B^\delta \\
&= (\varepsilon_{\alpha\delta}\, A^\alpha B^\delta)^2.
\end{aligned}$$

To remove the ambiguity in sign the convention is made that we choose that value of θ which satisfies

$$\sin\theta = +\, \varepsilon_{\alpha\delta}\, A^\alpha B^\delta. \tag{43.2}$$

In accordance with the convention just made, we say that the rotation from C^α to D^α is positive if the invariant $\varepsilon_{\alpha\beta} C^\alpha D^\beta$ is positive. This in effect chooses the positive rotation as that one which rotates C^α to D^α through an angle less than or equal to π.

Let us form the contravariant vector

$$(43.3) \qquad B^\beta = \varepsilon^{\alpha\beta} A_\alpha$$

from the covariant *unit* vector A_α. Its magnitude is given by

$$\begin{aligned}(B)^2 = a_{\alpha\beta}B^\alpha B^\beta &= a_{\alpha\beta}\,\varepsilon^{\gamma\alpha}\,\varepsilon^{\delta\beta} A_\gamma A_\delta \\ &= a_{22}(\varepsilon^{12})^2(A_1)^2 + 2a_{12}\,\varepsilon^{12}\,\varepsilon^{21} A_1 A_2 + a_{11}(\varepsilon^{21})^2(A_2)^2 \\ &= \frac{1}{a}\,\{a_{22}(A_1)^2 - 2a_{12} A_1 A_2 + a_{11}(A_2)^2\}.\end{aligned}$$

That is, in virtue of (42.9)

$$\begin{aligned}(B)^2 &= a^{11}(A_1)^2 + 2a^{12} A_1 A_2 + a^{22}(A_2)^2 \\ &= a^{\alpha\beta} A_\alpha A_\beta = (A)^2 = 1.\end{aligned}$$

Thus B^α is a unit vector. Further the angle θ between A^α and B^α satisfies

$$\sin\theta = \varepsilon_{\alpha\beta} A^\alpha B^\beta = \varepsilon^{\alpha\beta} A_\alpha B_\beta = B^\beta B_\beta = 1.$$

Therefore $\theta = \pi/2$ and by the above convention, equation (43.3) determines the unit vector B^α orthogonal to the unit vector A^α and oriented so that the rotation from A^α to B^α is positive.

We now apply (43.2) to calculate the angle ω between the coordinate curves. The unit tangent vectors to the u^1- and u^2-curves are $\dfrac{1}{\sqrt{a_{11}}}\,\delta_1^\alpha$ and $\dfrac{1}{\sqrt{a_{22}}}\,\delta_2^\alpha$ respectively. Hence the angle ω satisfies

$$(43.4) \quad \sin\omega = \frac{1}{\sqrt{a_{11}a_{22}}}\,\varepsilon_{\alpha\beta}\,\delta_1^\alpha\,\delta_2^\beta = \frac{1}{\sqrt{a_{11}a_{22}}}\,\varepsilon_{12} = \sqrt{\frac{a}{a_{11}a_{22}}}.$$

We also note from this equation that the rotation from the direction of a u^1- curve to a u^2-curve is always positive.

It is easy to deduce that the necessary and sufficient

condition that the coordinate curves are orthogonal everywhere on the surface is that a_{12} vanishes everywhere. We then say that the coordinates are **orthogonal curvilinear.**

§ 44. Surface covariant differentiation

We can form Christoffel symbols starting with the fundamental surface tensor $a_{\alpha\beta}$, and hence introduce covariant and intrinsic derivatives which are now surface tensors. So $A_{\alpha,\beta}$ will denote the covariant derivative of A_α with respect to u^β. Written in full

$$A_{\alpha,\beta} = \frac{\partial A_\alpha}{\partial u^\beta} - \begin{Bmatrix} \varepsilon \\ \alpha\beta \end{Bmatrix} A_\varepsilon .$$

There will be no confusion between the Christoffel symbols formed with the g_{ij} and those formed with the tensor $a_{\alpha\beta}$ since the Latin and Greek indices will clearly distinguish which symbol is meant. We can also form the Riemann-Christoffel surface tensor $R^{\alpha}_{.\beta\gamma\delta}$ from $a_{\alpha\beta}$ and the curvature surface tensor $R_{\varepsilon\beta\gamma\delta} = a_{\varepsilon\alpha} R^{\alpha}_{.\beta\gamma\delta}$.

Following the argument of section 26, we can show that the geodesics on the surface are the solutions of the differential equations

$$\text{(44.1)} \qquad \frac{\delta}{\delta s}\left(\frac{du^\alpha}{ds}\right) \equiv \frac{d^2 u^\alpha}{ds^2} + \begin{Bmatrix} \alpha \\ \beta\gamma \end{Bmatrix} \frac{du^\beta}{ds}\frac{du^\gamma}{ds} = 0.$$

Corresponding to (27.1) the geodesics also satisfy the equation

$$\text{(44.2)} \qquad a_{\alpha\beta}\frac{du^\alpha}{ds}\frac{du^\beta}{ds} = 1.$$

Thus, in practice, we need only consider one of the two equations (44.1) together with the first order equation (44.2).

As in section 28 we can introduce a system of geodesic coordinates on the surface so that at any particular point, called the pole, all the surface Christoffel symbols are zero. At the pole, covariant and intrinsic derivatives reduce respectively to the corresponding partial and total derivatives.

Again the theory of parallelism, outlined in section 29, applies to surface vectors. The vector-field A^α is said to be parallel along the curve $u^\alpha = u^\alpha(t)$ if

$$\frac{\delta A^\alpha}{\delta t} \equiv \frac{dA^\alpha}{dt} + \begin{Bmatrix} \alpha \\ \beta\gamma \end{Bmatrix} A^\beta \frac{du^\gamma}{dt} = 0.$$

Note that the covariant derivatives of $a_{\alpha\beta}$, $a^{\alpha\beta}$ and δ^α_β are zero. We cannot apply the method of section 38 to prove that $\varepsilon_{\alpha\beta,\gamma}$ and $\varepsilon^{\alpha\beta}_{,\gamma}$ are both zero because it is generally impossible to choose a rectangular cartesian system of coordinates on an arbitrary surface. Instead we select a geodesic coordinate system. At its pole, the Christoffel symbols are zero. Hence the partial derivatives of $a_{\alpha\beta}$ are also zero. Thus the partial derivatives of the determinant a are zero at the pole. That is $\varepsilon_{\alpha\beta,\gamma}$ and $\varepsilon^{\alpha\beta}_{,\gamma}$ are both zero there. It follows immediately that these tensors are zero at every point on the surface and in all coordinate systems. Hence the permutation surface tensors behave like constants with respect to surface covariant and intrinsic differentiation.

Ex. **Show that the conditions that the u^1-curves and the u^2-curves be geodesics are $\begin{Bmatrix} 2 \\ 11 \end{Bmatrix} = 0$ and $\begin{Bmatrix} 1 \\ 22 \end{Bmatrix} = 0$ respectively. Simplify these conditions if the coordinate system is orthogonal curvilinear.**

§ 45. Geodesic curvature

We discussed in section 40 the theory of the twisted curve in space. To this we now add the corresponding theory applicable to twisted curves lying on surfaces. Let the curve be given by the equations $u^\alpha = u^\alpha(s)$, where s measures arc-distance along the curve. The *unit* surface tangent vector is

$$(45.1) \qquad t^\alpha = \frac{du^\alpha}{ds}.$$

Since t^α is a unit vector, we have $a_{\alpha\beta}t^\alpha t^\beta = 1$ and intrinsic differentiation yields us $a_{\alpha\beta}t^\alpha \frac{\delta t^\beta}{\delta s} = 0$. This shows that $\delta t^\alpha/\delta s$ is a vector on the surface orthogonal to the tangent vector t^α. Let us denote the unit vector in the direction of $\delta t^\alpha/\delta s$ by n^α. Then

$$(45.2) \qquad \frac{\delta t^\alpha}{\delta s} = \sigma n^\alpha$$

where σ is an invariant called the **geodesic curvature** of the curve. We also call n^α the unit **normal vector** in the surface to the curve, and choose its direction so that the rotation from t^α to n^α is positive. Thus

$$\varepsilon_{\alpha\beta}t^\alpha n^\beta = 1,$$

and equations (45.2) determine σ uniquely in sign as well as magnitude. It is now clear from (43.3) and section 17 that

$$n^\beta = + \varepsilon^{\alpha\beta}t_\alpha$$

and

$$t^\beta = - \varepsilon^{\alpha\beta}n_\alpha.$$

Intrinsic differentiation of the first equation gives us

$$\frac{\delta n^\beta}{\delta s} = \varepsilon^{\alpha\beta}\frac{\delta t_\alpha}{\delta s} = \sigma\varepsilon^{\alpha\beta}n_\alpha = - \sigma t^\beta.$$

Combining this with (45.2) we have the surface **Frenet** equations of a curve

$$(45.3) \qquad \frac{\delta t^\alpha}{\delta s} = \sigma n^\alpha; \quad \frac{\delta n^\alpha}{\delta s} = - \sigma t^\alpha.$$

Although t^α is the same vector as T^i, their components being connected by the relation $T^i = x^i_\alpha t^\alpha$, it is important to note that n^α is in general neither the principal normal

N^i nor the binormal B^i. It does lie however in the normal plane of the curve determined by N^i and B^i.

Along a geodesic of the surface $\delta t^\alpha/\delta s = 0$ and hence $\sigma = 0$. Conversely if $\sigma = 0$, then $\delta t^\alpha/\delta s = 0$ and so the curve is a geodesic. Therefore, the necessary and sufficient condition that a curve be a geodesic is that the geodesic curvature be zero.

Ex. Prove that the geodesic curvatures of the coordinate curves are

$$\sigma_{(1)} = \sqrt{\frac{a}{(a_{11})^3}} \begin{Bmatrix} 2 \\ 11 \end{Bmatrix} \text{ and } \sigma_{(2)} = -\sqrt{\frac{a}{(a_{22})^3}} \begin{Bmatrix} 1 \\ 22 \end{Bmatrix}.$$

§ 46. Normal vector

We shall now derive an expression for the unit normal vector ξ^i at any point on a surface. Let us choose its orientation so that the u^1-curve, the u^2-curve and the normal at the point form a right-handed system. The surface unit tangent vectors to the coordinate curves are $\frac{1}{\sqrt{a_{11}}}\delta_1^\alpha$ and $\frac{1}{\sqrt{a_{22}}}\delta_2^\alpha$ respectively, and the corresponding space components are $\frac{1}{\sqrt{a_{11}}} x_\alpha^i \delta_1^\alpha$ and $\frac{1}{\sqrt{a_{22}}} x_\alpha^i \delta_2^\alpha$, that is $\frac{1}{\sqrt{a_{11}}} x_1^i$ and $\frac{1}{\sqrt{a_{22}}} x_2^i$. In virtue of (43.4) and the covariant form of (39.1) we have

$$\sqrt{\frac{a}{a_{11}a_{22}}}\, \xi_i = \varepsilon_{ijk} \frac{1}{\sqrt{a_{11}a_{22}}} x_1^j x_2^k.$$

That is,

$$\xi_i = \frac{1}{\sqrt{a}} \varepsilon_{ijk} x_1^j x_2^k. \tag{46.1}$$

It is not clear that ξ_i is a covariant vector because $\sqrt{a}$ appears in this equation. However the vector form is clearly seen from the equivalent equation

$$\xi_i = \tfrac{1}{2}\varepsilon^{\alpha\beta} \varepsilon_{ijk} x_\alpha^j x_\beta^k. \tag{46.2}$$

For purposes of calculation (46.1) is the more suitable and can be readily written as a determinant, but (46.2) is preferable for theoretical purposes. Since the vector ξ_i does not lie in the surface, there is no corresponding surface vector ξ_α. We shall often require the important equation

$$g_{ij}\xi^i x^j_\beta = 0, \tag{46.3}$$

which expresses the fact that the normal ξ^i is orthogonal to the surface vector x^i_β. This equation is also an immediate deduction from (46.1).

Ex. The **tangent surface** of a curve is defined to be the surface generated by all its tangent lines, called the rulings. Show that the normals to the surface along a ruling are all parallel to the binormal of the curve at the point of contact of the ruling.

§ 47. Tensor derivatives of tensors

In the theory of surfaces we require tensors which possess both Latin and Greek indices, for example x^i_α. All such tensors in this chapter will be contravariant with respect to space but covariant with respect to the surface. We can then select A^i_α as a typical tensor. When we change the coordinate systems both in space and on the surface the transformation law is

$$\bar{A}^i_\alpha = A^j_\beta \frac{\partial \bar{x}^i}{\partial x^j} \frac{\partial u^\beta}{\partial \bar{u}^\alpha}.$$

We now ask what tensors can be constructed by differentiation? Here we follow section 30 closely. As our space is Euclidean the concept of parallelism in it does not depend on the selection of a curve. Accordingly choose an arbitrary parallel vector-field X_i in space. On the surface take an arbitrary parallel vector-field Y^α along the curve C whose parameter is t. Then

$$\frac{dX_i}{dt} - \left\{ {j \atop ik} \right\} X_j \frac{dx^k}{dt} = 0$$

and

$$\frac{dY^\alpha}{dt} + \begin{Bmatrix} \alpha \\ \beta\gamma \end{Bmatrix} Y^\beta \frac{du^\gamma}{dt} = 0.$$

Form the invariant $A^i_\alpha X_i Y^\alpha$. By differentiation and the use of the equations of parallelism we have

$$\frac{d}{dt}\left[A^i_\alpha X_i Y^\alpha\right] = \frac{dA^i_\alpha}{dt} X_i Y^\alpha + A^i_\alpha \begin{Bmatrix} j \\ ik \end{Bmatrix} X_j \frac{dx^k}{dt} Y^\alpha - A^i_\alpha X_i \begin{Bmatrix} \alpha \\ \beta\gamma \end{Bmatrix} Y^\beta \frac{du^\gamma}{dt}.$$

That is, on making an appropriate change of dummy indices,

$$\frac{d}{dt}[A^i_\alpha X_i Y^\alpha] = \left[\frac{\partial A^i_\alpha}{\partial u^\beta} + \begin{Bmatrix} i \\ jk \end{Bmatrix} A^j_\alpha x^k_\beta - \begin{Bmatrix} \varepsilon \\ \alpha\beta \end{Bmatrix} A^i_\varepsilon\right] \frac{du^\beta}{dt} X_i Y^\alpha.$$

Applying the quotient law we see that the expression in square brackets is a tensor, which we call the **tensor derivative** of A^i_α with respect to u^β and we denote it by the semi-colon notation

$$(47.1) \qquad A^i_{\alpha;\beta} \equiv \frac{\partial A^i_\alpha}{\partial u^\beta} + \begin{Bmatrix} i \\ jk \end{Bmatrix} A^j_\alpha x^k_\beta - \begin{Bmatrix} \varepsilon \\ \alpha\beta \end{Bmatrix} A^i_\varepsilon.$$

It is possible to choose a rectangular cartesian system in space and a geodesic system on the surface. Then at the pole, the tensor derivatives are the partial derivatives. Consequently the laws of tensor differentiation are the same as those that apply to covariant derivatives.

We must now extend the concept of tensor differentiation further to space tensors and to surface tensors. It is clear that the tensor derivatives of surface tensors are identical with their covariant derivatives. Following the method of this section, we see that the tensor derivative of a space tensor with respect to u^α is the tensor obtained by inner multiplication of its covariant derivative with respect to x^l by the tensor x^l_α.

For example, $A^{ij}{}_{;\alpha} = A^{ij}{}_{,k} x^k_\alpha$. Thus the tensor derivatives of g_{ij}, g^{ij}, δ^i_j, ε_{ijk}, ε^{ijk}, $a_{\alpha\beta}$, $a^{\alpha\beta}$, δ^α_β, $\varepsilon_{\alpha\beta}$ and $\varepsilon^{\alpha\beta}$ are all zero. That is, they may be treated as constants with respect to tensor differentiation.

§ 48. Second fundamental form

By tensor differentiation

$$x^i_{\alpha;\beta} = \frac{\partial^2 x^i}{\partial u^\alpha \partial u^\beta} + \begin{Bmatrix} i \\ jk \end{Bmatrix} x^j_\alpha x^k_\beta - \begin{Bmatrix} \gamma \\ \alpha\beta \end{Bmatrix} x^i_\gamma, \tag{48.1}$$

which shows that $x^i_{\alpha;\beta}$ is symmetric in α and β. That is, $x^i_{\alpha;\beta} = x^i_{\beta;\alpha}$. Now the tensor derivation of (42.2) yields

$$g_{ij} x^i_{\alpha;\gamma} x^j_\beta + g_{ij} x^i_\alpha x^j_{\beta;\gamma} = 0.$$

We subtract this equation from the sum of the two similar equations obtained by cyclic interchange of α, β and γ. Because $x^i_{\alpha;\beta}$ is symmetric the result is

$$g_{ij} x^i_{\alpha;\beta} x^j_\gamma = 0. \tag{48.2}$$

This shows that $x^i_{\alpha;\beta}$ is a contravariant space vector orthogonal to all vectors x^j_γ lying on the surface. Thus it is co-directional with the normal vector ξ^i. Hence quantities $b_{\alpha\beta}$ must exist so that

$$x^i_{\alpha;\beta} = b_{\alpha\beta} \xi^i. \tag{48.3}$$

Further, it follows that $b_{\alpha\beta}$ form the components of a covariant symmetric surface tensor. Equations (48.3) are known as **Gauss's formulae.** Since ξ^i is a unit vector, inner multiplication of (48.3) by ξ_i and substitution from (46.2) yields

$$b_{\alpha\beta} = \tfrac{1}{2}\varepsilon^{\gamma\delta} \varepsilon_{ijk} x^i_{\alpha;\beta} x^j_\gamma x^k_\delta = \frac{1}{\sqrt{a}} \varepsilon_{ijk} x^i_{\alpha;\beta} x^j_1 x^k_2. \tag{48.4}$$

The quadratic form

$$b_{\alpha\beta}\, du^\alpha\, du^\beta \tag{48.5}$$

is called the **second fundamental form** of the surface. It is now possible to construct the invariant

(48.6) $$H = \tfrac{1}{2}a^{\alpha\beta} b_{\alpha\beta},$$

which is called the **mean curvature** of the surface.

Ex. If the space coordinates are rectangular Cartesian, show that

$$b_{\alpha\beta} = \frac{1}{\sqrt{a}} \varepsilon_{ijk} \frac{\partial^2 x^i}{\partial u^\alpha \partial u^\beta} x^j_1 x^k_2.$$

§ 49. Third fundamental form

Tensor differentiation of the identity $g_{ij}\xi^i\xi^j = 1$ yields us $g_{ij}\xi^i\xi^j_{;\alpha} = 0$. That is, $\xi^j_{;\alpha}$ is a contravariant space vector orthogonal to the normal vector. Accordingly it lies in the surface. Hence quantities η^β_α exist so that

(49.1) $$\xi^i_{;\alpha} = \eta^\beta_\alpha x^i_\beta.$$

The quotient law then states that η^β_α form a mixed surface tensor. We now differentiate (46.3) tensorially and obtain

$$g_{ij}\xi^i_{;\alpha} x^j_\beta + g_{ij}\xi^i x^j_{\beta;\alpha} = 0,$$

which reduces on substitution from (49.1) and (48.3) to

$$g_{ij}\eta^\gamma_\alpha x^i_\gamma x^j_\beta + g_{ij}\xi^i b_{\alpha\beta}\xi^j = 0.$$

We apply (42.2) and the result is

(49.2) $$b_{\alpha\beta} = - a_{\beta\gamma}\eta^\gamma_\alpha.$$

Inner multiplication by $a^{\beta\epsilon}$ yields

(49.3) $$\eta^\epsilon_\alpha = - a^{\beta\epsilon} b_{\alpha\beta};$$

and so we can rewrite (49.1) in the form

(49.4) $$\xi^i_{;\alpha} = - a^{\gamma\beta} b_{\alpha\gamma} x^i_\beta.$$

These equations are known as **Weingarten's formulae.** Introduce the symmetric surface tensor

(49.5) $$c_{\alpha\beta} = g_{ij}\xi^i_{;\alpha}\xi^j_{;\beta}$$

and form from it the quadratic form $c_{\alpha\beta}du^\alpha du^\beta$ which is then called the **third fundamental form** of the surface.

Ex. Prove that $c_{\alpha\beta} = a^{\gamma\delta} b_{\alpha\gamma} b_{\beta\delta}$.

§ 50. Gauss-Codazzi equations

We are now in a position to obtain the central formulae in the theory of surfaces. For the moment, choose the space coordinates to be rectangular cartesian and the surface coordinates to be geodesic. Then at the *pole*, tensor differentiation of (48.1) yields

$$x^i_{\alpha;\beta\gamma} = \frac{\partial^3 x^i}{\partial u^\alpha \partial u^\beta \partial u^\gamma} - \frac{\partial}{\partial u^\gamma}\begin{Bmatrix}\sigma\\ \alpha\beta\end{Bmatrix} x^i_\sigma.$$

Thus

$$x^i_{\alpha;\beta\gamma} - x^i_{\alpha;\gamma\beta} = \left[\frac{\partial}{\partial u^\beta}\begin{Bmatrix}\sigma\\ \alpha\gamma\end{Bmatrix} - \frac{\partial}{\partial u^\gamma}\begin{Bmatrix}\sigma\\ \alpha\beta\end{Bmatrix}\right] x^i_\sigma.$$

The expression in square brackets is the surface Riemann-Christoffel tensor $R^\sigma_{.\alpha\beta\gamma}$ at the pole. Therefore we have set up a tensor equation

$$x^i_{\alpha;\beta\gamma} - x^i_{\alpha;\gamma\beta} = R^\sigma_{.\alpha\beta\gamma} x^i_\sigma \tag{50.1}$$

which must be true at every point of the surface and in all coordinate systems.

We substitute from (49.4) in the tensor derivative of (48.3) and obtain

$$x^i_{\alpha;\beta\gamma} = b_{\alpha\beta;\gamma}\xi^i - a^{\epsilon\sigma} b_{\alpha\beta} b_{\gamma\epsilon} x^i_\sigma.$$

We can consequently write (50.1) in the form

$$(b_{\alpha\beta;\gamma} - b_{\alpha\gamma;\beta})\xi^i - a^{\epsilon\sigma}(b_{\alpha\beta} b_{\gamma\epsilon} - b_{\alpha\gamma} b_{\beta\epsilon}) x^i_\sigma = R^\sigma_{.\alpha\beta\gamma} x^i_\sigma.$$

Inner multiplication by ξ_i and $g_{ij} x^j_\rho$ in virtue of (42.2) and (46.3) yield the respective equations

$$b_{\alpha\beta;\gamma} - b_{\alpha\gamma;\beta} = 0 \tag{50.2}$$

and

$$R_{\rho\alpha\beta\gamma} = b_{\alpha\gamma} b_{\rho\beta} - b_{\alpha\beta} b_{\rho\gamma}. \tag{50.3}$$

The reader is asked to verify that (50.2) consists of only two independent partial differential equations. They are called the **Codazzi** equations. As there is only one distinct component of the curvature tensor in two dimensions, (50.3) reduces to the single equation

$$R_{1212} = b_{11}b_{22} - (b_{12})^2 \tag{50.4}$$

which is called the **Gauss** equation. By means of (35.2) we can write this equation

$$K = \frac{b}{a} \tag{50.5}$$

where b is the determinant formed from the $b_{\alpha\beta}$ and K is the Riemannian curvature of the surface. On a surface it is more usual to call K the **Gaussian** or **total curvature.**

It can be proved* that a surface is uniquely determined except for a translation or rotation in space when the first and second fundamental forms are given. This theorem can be precisely formulated as follows;- If $a_{\alpha\beta}$ and $b_{\alpha\beta}$ are given functions of u^1 and u^2, then there exists a surface $x^i = x^i(u^\alpha)$, uniquely determined except for its position in space, which has $a_{\alpha\beta}du^\alpha du^\beta$ and $b_{\alpha\beta}du^\alpha du^\beta$ as its first and second fundamental forms respectively, provided that $a_{\alpha\beta}du^\alpha du^\beta$ is positive-definite and that $a_{\alpha\beta}$ and $b_{\alpha\beta}$ satisfy the Gauss-Codazzi equations.

Ex. Prove that $c_{\alpha\beta} = 2Hb_{\alpha\beta} - Ka_{\alpha\beta}$.

§ 51. Normal curvature-asymptotic lines

On a surface consider the curve $u^\alpha = u^\alpha(s)$, where s measures arc-distance. The equations of the curve in space will be $x^i = x^i(s)$. Then the space vectors T^i, N^i and B^i and the surface vectors t^α and n^α at any point of the curve satisfy the Frenet formulae (40.5) and (45.3).

* L. P. Eisenhart, Differential Geometry, p. 157—159.

The tangent vectors T^i and t^α are connected by $T^i = t^\alpha x^i_\alpha$. Intrinsic differentiation yields in virtue of (48.3)

$$\frac{\delta T^i}{\delta s} = T^i_{,j}\frac{dx^j}{ds} = T^i_{,j}x^j_\beta\frac{du^\beta}{ds} = T^i_{;\beta}\frac{du^\beta}{ds}$$

$$= t^\alpha_{;\beta}x^i_\alpha\frac{du^\beta}{ds} + t^\alpha x^i_{\alpha;\beta}\frac{du^\beta}{ds}$$

$$= \frac{\delta t^\alpha}{\delta s}x^i_\alpha + b_{\alpha\beta}t^\alpha t^\beta \xi^i.$$

Applying the Frenet formulae we have

$$\kappa N^i = \sigma n^\alpha x^i_\alpha + b_{\alpha\beta}t^\alpha t^\beta \xi^i,$$

which becomes

$$\kappa N^i = \sigma n^i + b_{\alpha\beta}t^\alpha t^\beta \xi^i \tag{51.1}$$

when we designate the space components of n^α by n^i. Let us introduce the angle θ between the principal normal N^i and the surface normal ξ^i. Then inner multiplication of (51.1) by ξ_i yields

$$\kappa \cos\theta = b_{\alpha\beta}t^\alpha t^\beta, \tag{51.2}$$

since ξ_i is orthogonal to the vector n^i which lies in the surface. The invariant $b_{\alpha\beta}t^\alpha t^\beta$ is the same for all curves which have the same tangent vector t^α at the point on the surface. Accordingly we deduce **Meusnier's theorem** 'For all curves on a surface which have the same tangent vector, the quantity $\kappa\cos\theta$ is constant'. This quantity is called the **normal curvature** at the point and is denoted by $\kappa_{(n)}$. Hence

$$\kappa_{(n)} = b_{\alpha\beta}t^\alpha t^\beta = \frac{b_{\alpha\beta}\,du^\alpha du^\beta}{a_{\alpha\beta}\,du^\alpha du^\beta}. \tag{51.3}$$

If we choose a plane section through the normal to the surface, then $\theta = 0$ or π. That is $\kappa_{(n)} = \pm\kappa$. So the normal curvature in any direction is equal in magnitude to the curvature of the normal plane section of the surface in that direction.

Along a geodesic $\sigma = 0$, therefore equations (51.1) reduce to $\kappa N^i = b_{\alpha\beta} t^\alpha t^\beta \xi^i$. Thus either $\kappa = 0$ or $N^i = \pm \xi^i$. We deduce that a geodesic on a surface is either a straight line or is a curve whose principal normal is co-directional with the surface normal at every point. Conversely, if $N^i = \pm \xi^i$, then inner multiplication of (51.1) by n_i yields $\sigma = 0$. That is, the curve is a geodesic.

The directions at a point on a surface which satisfy the equation $b_{\alpha\beta} du^\alpha du^\beta = 0$ are called the **asymptotic directions.** If all the points of a curve have asymptotic tangent directions, the curve is called an **asymptotic line.** The asymptotic lines on a surface are given by $b_{\alpha\beta} du^\alpha du^\beta = 0$. Thus along an asymptotic line we have $\kappa N^i = \sigma n^i$. Hence, since N^i and n^i are both unit vectors, either $\kappa = \sigma = 0$ or the curvature and geodesic curvature of an asymptotic line are equal in magnitude and its principal normal lies in the surface. Consequently the binormal of an asymptotic line, which is not a straight line, is co-directional with the surface normal. The converse is also valid.

Ex. Prove Enneper's formula that the torsion τ of an asymptotic line is $\pm \sqrt{-K}$ where K is the Gaussian curvature of the surface.

§ 52. Principal curvatures - lines of curvature

The normal curvature $\kappa_{(n)}$ of a surface in the direction t^α is given by $\kappa_{(n)} = b_{\alpha\beta} t^\alpha t^\beta$, where the tangent vector satisfies $a_{\alpha\beta} t^\alpha t^\beta = 1$. The maximum and minimum values of $\kappa_{(n)}$ can then be determined by the method used in section 19. They correspond to the principal directions determined by $b_{\alpha\beta}$ and are given by the roots of the determinantal equation $| b_{\alpha\beta} - \lambda a_{\alpha\beta} | = 0$, which reduces in virtue of (50.5) and (48.6) to

$$\lambda^2 - 2H\lambda + K = 0. \tag{52.1}$$

The roots $\kappa_{(1)}$ and $\kappa_{(2)}$ of this equation are called the **principal curvatures** of the surface at the point. The principal directions $t^\alpha_{(1)}$ and $t^\alpha_{(2)}$ corresponding to the principal curvatures at the point respectively satisfy

$$(b_{\alpha\beta} - \kappa_{(1)} a_{\alpha\beta})t^\beta_{(1)} = 0,$$
$$(b_{\alpha\beta} - \kappa_{(2)} a_{\alpha\beta})t^\beta_{(2)} = 0.$$

A point at which $\kappa_{(1)} = \kappa_{(2)}$ is called an **umbilic.** At all other points section 19 tells us that $t^\alpha_{(1)}$ and $t^\alpha_{(2)}$ are orthogonal to one another. A curve on the surface which is a principal direction at all of its points is called a **line of curvature.**

At an umbilic, equation (52.1) has coincident roots. That is $H^2 = K$ and the reader may verify* that this result can be written

$$4a(a_{11}b_{12} - a_{12}b_{11})^2$$
$$+ [a_{11}(a_{11}b_{22} - a_{22}b_{11}) - 2a_{12}(a_{11}b_{12} - a_{12}b_{11})]^2 = 0.$$

Since $a_{\alpha\beta}du^\alpha du^\beta$ is positive-definite, a is positive, and we deduce that

$$a_{11}b_{12} - a_{12}b_{11} = a_{11}b_{22} - a_{22}b_{11} = 0,$$

and so

$$\frac{b_{11}}{a_{11}} = \frac{b_{12}}{a_{12}} = \frac{b_{22}}{a_{22}}.$$

Equation (51.3) now shows that $\kappa_{(n)}$ is independent of the direction du^α/ds. That is, at an umbilic, the normal curvature is the same in every direction**.

Ex. 1. Prove that the lines of curvature on a surface are given by $\varepsilon^{\gamma\delta} a_{\alpha\gamma} b_{\beta\delta} du^\alpha du^\beta = 0$.

Ex. 2. If the coordinate curves are lines of curvature, prove that $a_{12} = b_{12} = 0$, and conversely.

Ex. 3. Prove that a surface, all of whose points are umbilics, is a sphere or a plane.

* L. P. Eisenhart, Differential Geometry, p. 119.

** The reader may find it interesting to relate some of the results of this chapter to chapter II of Rutherford's Vector Methods.

CHAPTER VII

CARTESIAN TENSORS – ELASTICITY

§ 53. Orthogonal transformations

In this chapter our aim is to present the Theory of Elasticity and so we restrict ourselves to a Euclidean space of three dimensions. In it we choose a right-handed system of rectangular cartesian coordinates and we denote them by y_i, where Latin indices have the range 1 to 3. The line-element ds is then given by

$$ds^2 = dy_i\, dy_i = \delta_{ij}\, dy_i\, dy_j, \tag{53.1}$$

where δ_{ij} is the Kronecker delta.

The linear equations, (compare (6.3)),

$$\bar{y}_i = a_{ij} y_j + b_i, \tag{53.2}$$

where b_i form a set of three constants and a_{ij} a set of nine constants define a transformation to a new coordinate system $\bar{y}_i$. The necessary and sufficient conditions that the $\bar{y}_i$ form a set of rectangular cartesian coordinates is

$$ds^2 = d\bar{y}_i\, d\bar{y}_i = a_{ij}\, a_{ik}\, dy_j\, dy_k = \delta_{jk}\, dy_j\, dy_k.$$

That is,

$$(a_{ij}\, a_{ik} - \delta_{jk}) dy_j\, dy_k = 0$$

for all values of dy_j. Hence

$$a_{ij}\, a_{ik} = \delta_{jk}. \tag{53.3}$$

Inner multiplication of (53.2) by a_{ik} yields the solution

$$y_k = a_{ik} \bar{y}_i - a_{ik} b_i. \tag{53.4}$$

Thus

$$\frac{\partial \bar{y}_i}{\partial y_j} = \frac{\partial y_j}{\partial \bar{y}_i} = a_{ij}. \tag{53.5}$$

In virtue of these equations, we see by examining (9.1) that the distinction between contravariance and covariance has disappeared. Accordingly we shall write all indices as subscripts on condition that we allow only transformations of the type (53.2) subject to (53.3). We have already anticipated this in our notation y_i, δ_{ij} and a_{ij}. However we may sometimes wish to adopt a curvilinear coordinate system such as spherical polars. It is then necessary to reintroduce the distinction between contravariance and covariance. This will be indicated by a return to the coordinates x^i.

The transformation (53.2) is equivalent to the combination of the two transformations $\bar{y}_i = y'_i + b_i$ and $y'_i = a_{ij} y_j$. The transformation $\bar{y}_i = y'_i + b_i$ merely defines a translation to new parallel axes. The transformation

$$y'_i = a_{ij} y_j \tag{53.6}$$

subject to the six conditions (53.3) is said to define an **orthogonal transformation.** It follows from (53.3) that the determinant $| a_{ij} |$ of an orthogonal transformation is either $+1$ or -1 and we say that (53.6) defines a positive or a negative orthogonal transformation respectively. It is well known* that the y'_i axes form a right-handed or a left-handed system if the orthogonal transformation is positive or negative respectively. Further a positive orthogonal transformation defines a rotation of the axes about the origin.

An alternative set of equations to (53.3) is obtained by considering

* W. H. McCrea, 'Analytical Geometry of Three Dimensions', pp. 10—13.

$$ds^2 = dy_i dy_i = a_{ji} a_{ki} d\bar{y}_j d\bar{y}_k = \delta_{jk} d\bar{y}_j d\bar{y}_k,$$

from which we deduce that

$$a_{ji} a_{ki} = \delta_{jk}. \tag{53.7}$$

Ex. Show that in the rotation of axes defined by the positive orthogonal transformation (53.6), a_{ij} is the cosine of the angle between the y'_i axis and the y_j axis. The components a_{ij} are therefore the direction cosines of the y'_i system with respect to the y_j system.

§ 54. Rotations

Equations (53.2) may be interpreted from another point of view. We could say that they transform the point P whose coordinates are y_i into the point $\bar{P}$ whose coordinates are $\bar{y}_i$ referred to the *same* system of rectangular cartesian axes. We then call (53.2) an **affine transformation***. If in addition conditions (53.3) together with $| a_{ij} | = 1$ are imposed, we have the most general rigid body motion consisting of a rotation followed by a translation.

Next we wish to obtain the orthogonal transformation which represents a right-handed rotation through an angle ψ about a line through the origin whose direction

* **Affine Geometry.** According to Klein's Erlangen Programm, (F. Klein, Math. Ann. Vol 43, (1893), p. 63), a geometry comprises a system of definitions and theorems that express properties which are invariant with respect to a given group of transformations. For example, if the group of transformations consists of all rigid body motions, namely translations and rotations, then the geometry is termed **metrical.** This is the geometry of Euclid, less the similarity theorems, and its most important concepts are *distance* and *angle.*

Let us now examine **affine** geometry, defined by the group of transformations (53.2). Suppose A_i to be the vector joining the point with coordinates z_i to the point with coordinates y_i. Then $A_i = y_i - z_i$, and we immediately deduce from (53.2) that the transformation law of vectors is $\bar{A}_i = a_{ij} A_j$. We now choose two parallel vectors A_i and B_i. The requisite conditions are

cosines are l_i. By a right-handed rotation, we mean that a right-handed corkscrew will move outwards from the origin along the direction l_i when it is rotated through an angle ψ about this direction. Using the vector notation we have (fig.)

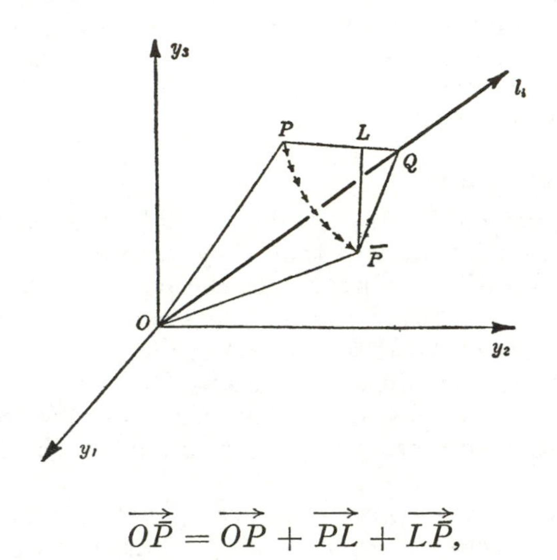

$$\overrightarrow{O\bar{P}} = \overrightarrow{OP} + \overrightarrow{PL} + \overrightarrow{L\bar{P}},$$

where Q is the perpendicular from P on the line through the origin whose direction cosines are l_i, the angle

---

$A_1/B_1 = A_2/B_2 = A_3/B_3$. But each fraction equals $(a_{11}A_1 + a_{12}A_2 + a_{13}A_3)/(a_{11}B_1 + a_{12}B_2 + a_{13}B_3) = \bar{A}_1/\bar{B}_1$, and similarly equals $\bar{A}_2/\bar{B}_2$ and $\bar{A}_3/\bar{B}_3$. Thus we have $\bar{A}_1/\bar{B}_1 = \bar{A}_2/\bar{B}_2 = \bar{A}_3/\bar{B}_3$. That is, the vectors $\bar{A}_i$ and $\bar{B}_i$ are parallel. Hence the *parallelism of vectors* is invariant in affine geometry.

Further we see from (53.2) that every finite point transforms into a finite point. Therefore the plane at infinity is invariant, and so we can distinguish between a non-central and a central quadric according as the plane at infinity does or does not touch the quadric. However we cannot define either *angle* or *distance* in affine geometry, because they are not invariants under the affine group of transformations (53.2).

$P Q \bar{P} = \psi$, $QP = Q\bar{P}$, the plane $PQ\bar{P}$ is orthogonal to l_i and $\bar{P}L$ is the perpendicular from $\bar{P}$ to PQ. Since $\overrightarrow{OP} = y_i$, we have $\overrightarrow{OQ} = l_i l_k y_k$ and thus $\overrightarrow{PQ} = l_i l_k y_k - y_i$. Therefore $\overrightarrow{PL} = (1 - \cos\psi)\overrightarrow{PQ} = (1 - \cos\psi)(l_i l_k y_k - y_i)$. Further $\overrightarrow{L\bar{P}}$ is orthogonal to both $\overrightarrow{PQ}$ and l_i in such a way that $\overrightarrow{PQ}$, l_i and $\overrightarrow{L\bar{P}}$ form a right-handed system. Then by (39.1) the *unit* vector in the direction $\overrightarrow{L\bar{P}}$ is given by $e_{ijk}(l_j l_m y_m - y_j) l_k / PQ = -e_{ijk} y_j l_k / PQ$. Hence $\overrightarrow{L\bar{P}} = PQ \sin\psi(-e_{ijk} y_j l_k / PQ) = e_{ijk} l_j y_k \sin\psi$. Thus we have

$$\bar{y}_i = y_i + (1 - \cos\psi)(l_i l_k y_k - y_i) + \sin\psi\, e_{ijk} l_j y_k,$$

which can be written

$$\bar{y}_i = a_{ik} y_k \tag{54.1}$$

where

$$a_{ik} = \cos\psi\, \delta_{ik} + (1 - \cos\psi) l_i l_k + \sin\psi e_{ijk} l_j. \tag{54.2}$$

In the theory of elasticity we shall be particularly interested in infinitesimal rotations, in which case $\cos\psi \doteq 1$ and $\sin\psi \doteq \psi$. The infinitesimal rotation is then represented by

$$\bar{y}_i = y_i + \psi e_{ijk} l_j y_k.$$

That is,

$$\bar{y}_i = y_i + s_{ik} y_k, \tag{54.3}$$

where

$$s_{ik} = \psi e_{ijk} l_j. \tag{54.4}$$

It is clear that s_{ik} is anti-symmetric. Conversely, let us consider equations (54.3) given that s_{ik} is anti-symmetric. Then equations (54.4) comprise only three equations for the three unknowns l_i, whose solution is in fact $l_1 = -s_{23}/\psi$, $l_2 = -s_{31}/\psi$ and $l_3 = -s_{12}/\psi$. Therefore equations (54.3) always represent an infinitesimal rotation if ψ is infinitesimal and s_{ik} is anti-symmetric.

Ex. Calculate a_{ik} corresponding to a rotation of 90° about the y_3 axis.

§ 55. Cartesian tensors

A **Cartesian tensor** of the M-th order in a three-dimensional Euclidean space is defined as a set of 3^M quantities which transform according to equations (9.1) when the coordinates undergo a positive orthogonal transformation. This is a less stringent condition than that imposed on a tensor. So we see that all tensors are Cartesian tensors but a Cartesian tensor is not necessarily a tensor in the usual sense. In virtue of (53.5) we have that $A_{k_1k_2..k_M}$ is a Cartesian tensor of the M-th order if the transformed components satisfy

$$(55.1) \qquad \bar{A}_{l_1l_2..l_M} = a_{l_1k_1}a_{l_2k_2}..a_{l_Mk_M}A_{k_1k_2..k_M},$$

on change of the coordinates by the positive orthogonal transformation $\bar{y}_i = a_{ij}y_j$. We see from (53.6) that both y_i and their differentials dy_i are Cartesian vectors. Also the Kronecker delta is a Cartesian tensor of the second order because

$$\bar{\delta}_{ij} = a_{ir}a_{js}\delta_{rs} = a_{ir}a_{jr} = \delta_{ij}$$

in virtue of (53.7). Further we deduce from (38.2) that the permutation symbol e_{ijk} is a Cartesian tensor of the third order. Thus a_{ik} and s_{ik} introduced in the last section are Cartesian tensors of the second order.

The fundamental tensor of the Euclidean space is the Kronecker delta δ_{ij}. Hence all the Christoffel symbols are zero and so the comma notation for covariant derivatives now denotes the familiar partial derivatives which are Cartesian tensors.

The use of (55.1) instead of (9.1) shows us clearly that the quotient law of section 12 still applies to Cartesian tensors.

§ 56. Infinitesimal strain

Consider a body which is strained by the action of applied forces. The particle at the point P with coordinates y_i referred to a rectangular cartesian system is displaced to the point $\bar{P}$ with coordinates $y_i + u_i$. Similarly the particle which was at Q with coordinates z_i is displaced to the point $\bar{Q}$ with coordinates $z_i + v_i$. We define the **extension** $e_{(PQ)}$ of the straight line joining the unstrained points P and Q to be the increase in length per unit of length due to the strain. That is

$$e_{(PQ)} = \frac{\bar{P}\bar{Q} - PQ}{PQ} = \frac{\bar{P}\bar{Q}}{PQ} - 1. \tag{56.1}$$

We have

$$(PQ)^2 = (y_i - z_i)(y_i - z_i),$$

and

$$\begin{aligned}(\bar{P}\bar{Q})^2 &= (y_i + u_i - z_i - v_i)(y_i + u_i - z_i - v_i)\\ &= (y_i - z_i)(y_i - z_i) + 2(y_i - z_i)(u_i - v_i)\\ &\qquad + (u_i - v_i)(u_i - v_i)\\ &= (PQ)^2\left\{1 - \frac{2l_i(u_i - v_i)}{PQ} + \frac{(u_i - v_i)(u_i - v_i)}{(PQ)^2}\right\}\end{aligned}$$

where $l_i = (z_i - y_i)/PQ$ are the direction cosines of the unstrained line PQ. Hence

$$e_{(PQ)} = \left\{1 - \frac{2l_i(u_i - v_i)}{PQ} + \frac{(u_i - v_i)(u_i - v_i)}{(PQ)^2}\right\}^{\frac{1}{2}} - 1.$$

We confine our attention to the case of infinitesimal strain, where the assumption is made that the displacement vectors u_i and v_i are small compared to PQ. Neglecting quantities of higher order than the first, we obtain

$$e_{(PQ)} = -\frac{l_i(u_i - v_i)}{PQ}.$$

Let us now choose Q to be in the neighbourhood of

P so that $y_i - z_i$ is small. Then by Taylor's theorem for a function of three variables*

$$v_i = u_i + (z_j - y_j)u_{i,j} + \text{terms of higher order in } z_j - y_j.$$

Neglecting terms of higher order than the first in $z_j - y_j$, we obtain that the extension e at P in the direction determined by the unit vector l_i is given by

$$e = \frac{l_i(z_j - y_j)}{PQ} u_{i,j} = u_{i,j} l_i l_j.$$

Introduce the symmetric Cartesian **strain tensor** e_{ij} by the equations

$$e_{ij} = \tfrac{1}{2}(u_{i,j} + u_{j,i}), \tag{56.2}$$

(there is no confusion between e_{ij} and the permutation symbols $e_{\alpha\beta}$ of the previous chapter which are distinguished by Greek indices), and we finally obtain the extension e as the quadratic form

$$e = e_{ij} l_i l_j. \tag{56.3}$$

The **dilatation** or **expansion** θ is defined as the increase in volume per unit of volume. That is

$$\theta = \frac{\Delta\bar{V} - \Delta V}{\Delta V} = \frac{\Delta\bar{V}}{\Delta V} - 1, \tag{56.4}$$

where $\Delta\bar{V}$ denotes the strained volume corresponding to the volume ΔV. But

$$\frac{\Delta\bar{V}}{\Delta V} = \left|\frac{\partial(y_i + u_i)}{\partial(y_j)}\right| = 1 + u_{1,1} + u_{2,2} + u_{3,3} + \text{terms of higher order.}$$

Thus in virtue of (56.2) and (56.4) we have

$$\theta = e_{ii}. \tag{56.5}$$

The components of the strain tensor are not entirely arbitrary. To prove this, differentiate (56.2) twice and obtain

* R. P. Gillespie, Partial Differentiation, p. 60.

$$e_{ij,kl} = \tfrac{1}{2}(u_{i,jkl} + u_{j,ikl}),$$

from which it immediately follows that

$$e_{ij,kl} + e_{kl,ij} - e_{ik,jl} - e_{jl,ik} = 0. \tag{56.6}$$

There appear to be 81 of these **compatibility equations.** In actual fact some are repeated due to the symmetry of the strain tensor and others are satisfied identically. The reader is asked to verify that only six of these equations are independent.

To conclude this section, we investigate several important examples of strain.

(1) **Uniform dilatation.** Consider the displacement vector $u_i = cy_i$ where c is a constant. We have $e_{ij} = c\delta_{ij}$ and the dilatation $\theta = 3c$. Thus the extension at any point in any direction is constant and equal to one-third of the dilatation.

(2) **Simple extension.** Consider the displacement vector $u_i = cl_i l_j y_j$ where c is a constant and l_i a *unit* vector. We have $u_{i,j} = cl_i l_j$ and so $e_{ij} = cl_i l_j$ and the dilatation $\theta = cl_i l_i = c$. Also $e_{ij} l_i l_j = c$ and thus there is an extension at any point in the direction l_i of amount equal to the dilatation. The extension in any direction orthogonal to l_i is easily seen to be zero. If c is negative we refer to it as **simple contraction.**

(3) **Shearing strain.** Consider the displacement vector $u_i = 2cl_i m_j y_j$ where c is constant and both l_i and m_i are *unit* vectors. A brief calculation yields $e_{ij} = c(l_i m_j + l_j m_i)$ and the dilatation $\theta = 2cl_i m_i$. Consequently the dilatation is zero if the directions l_i and m_i are orthogonal.

Ex. Show that a simple extension along any direction together with an equal simple contraction along an orthogonal direction is equivalent to a shearing strain along a direction bisecting the angle between the given directions.

§ 57. Stress

The forces acting on a body are either external or internal. The external forces may consist either of body forces such as gravity which act on every particle of it, or of surface forces which act on the external surface of the body, for example the pressure between two bodies in contact. If F_j denotes the body force vector per unit volume, then the force acting on an element of volume ΔV is $F_j \Delta V$. Similarly if T_j denotes the surface force vector per unit area, then the force acting on an element of surface ΔS is $T_j \Delta S$. In order to discuss the internal forces, we select a small element of area ΔS inside the body and denote the direction cosines of the normal to this element, which is approximately planar, by n_j. We call one side of the element ΔS positive and the other side negative. Then the action of the positive side on the negative side is the internal surface force $T_j \Delta S$ where T_j is the force per unit area on the element ΔS. It is called the **stress vector** and is in general a function of the coordinates of the point which determines the position of the element ΔS and of the direction cosines n_j of the normal to ΔS. It is well to emphasise that T_j is not necessarily co-directional with n_j. At all points on the external surface of the body T_j becomes the external surface force.

Consider a *small* rectangular parallelepiped with vertex at the point P whose edges are parallel to the coordinate axes. We form three stress vectors $T_{(1)j}$, $T_{(2)j}$ and $T_{(3)j}$ corresponding to the small elements of area through P which are parallel to the coordinate planes. The stress vector $T_{(i)j}$ will be called positive if it acts in the positive direction of the y_i axis provided that the external normal is co-directional with the positive y_i axis. If, however, the external normal is co-directional with the negative y_i axis, then the stress vector $T_{(i)j}$ is positive if it acts in the direction of the negative y_i axis. In other words, a stress which tends to stretch will be regarded as positive, whilst a stress which tends to compress is regarded as negative. We define nine quantities E_{ij} by the equations

$$E_{ij} = T_{(i)j} \tag{57.1}$$

and we shall show that E_{ij} is a Cartesian tensor, called the stress tensor.

Construct the small tetrahedron $PA_1A_2A_3$ such that the edges PA_i are parallel to the y_i axes. The forces acting on the tetrahedron are the body force $F_j\Delta V$, the surface forces $E_{ij}\,\Delta S_i$ (no summation over i) on the faces opposite to A_i and the surface force $T_j\Delta S$ acting on the face $A_1A_2A_3$, where ΔV is the volume of the tetrahedron, ΔS_i is the area of the face opposite to A_i and ΔS is the area of the face $A_1A_2A_3$. Let the positive direction of T_j be that of the normal drawn outwards from the tetrahedron and whose direction cosines are n_j and let p be the perpendicular distance from P to the face $A_1A_2A_3$. Then $\Delta V = \frac{1}{3}p\Delta S$ and $\Delta S_i = n_i\Delta S$. The equations of equilibrium obtained by resolving forces parallel to the axes are now

$$F_j\Delta V - E_{ij}\Delta S_i + T_j\Delta S = 0,$$

where the E_{ij} occurs with a negative sign, because the external normals are in the directions of the negative axes. We now substitute for ΔV and ΔS_i, then divide by ΔS and proceed to the limit as the tetrahedron shrinks to zero, in which case p tends to zero. The result is

$$T_j = E_{ij}n_i. \tag{57.2}$$

It follows from the quotient law that E_{ij} is a Cartesian tensor, and we can by (57.2) calculate the stress vector at any point, corresponding to any direction at that point in terms of its direction cosines and the stress tensor.

We now cite several important cases of stress:-

(1) **Normal Stress.** The vector T_j is co-directional with $\pm n_j$. We see from (57.2) that $E_{ij} = C\delta_{ij}$ where C

is a constant. Hydrostatic pressure is an example of normal stress for which C is negative.

(2) **Simple Tension.** Consider the stress tensor $E_{ij} = Cl_i l_j$ where C is a constant and l_i is a *unit* vector. Then the stress vector in the direction l_i is $T_j = Cl_i l_j l_i = Cl_j$ and it is thus co-directional with $\pm l_j$. However the stress vector in the direction m_i orthogonal to l_i is $T_j = Cl_i l_j m_i = 0$. If C is negative the stress is called a **Simple Compression.**

(3) **Shearing Stress.** This is specified by the stress tensor $E_{ij} = C(l_i m_j + l_j m_i)$ where C is a constant and l_i and m_i are *unit* vectors.

Ex. Show that a simple tension along any direction together with an equal simple compression along an orthogonal direction is equivalent to a shearing stress along a direction bisecting the angle between the given directions.

§ 58. Equations of equilibrium

Let us consider a body of volume V in equilibrium, which is enclosed by the surface S. We resolve forces parallel to the axes and obtain

$$\int_V F_j dV + \int_S T_j dS = 0.$$

Substitution from (57.2) yields

$$\int_V F_j dV + \int_S E_{ij} n_i dS = 0,$$

which on application of Gauss's theorem* becomes

$$\int_V F_j dV + \int_V E_{ij,i} dV = 0.$$

That is,

$$\int_V (F_j + E_{ij,i}) dV = 0.$$

This equation is an identity, being true for any volume V. Hence

$$F_j + E_{ij,i} = 0. \tag{58.1}$$

* D. E. Rutherford, Vector Methods, p. 74.

The moments of a force about the axes are the components of the vector product of the force vector and the position vector of any point on the line of action of the force. Thus in tensor notation, the moments of the force F_i about the axes are represented by the vector $e_{ijk}y_j F_k$. Now take moments about the axes for our body in equilibrium and the result is

$$\int_V e_{ijk}y_j F_k dV + \int_S e_{ijk}y_j T_k dS = 0.$$

We substitute from (57.2), apply Gauss's theorem and obtain

$$\int_V e_{ijk}y_j F_k dV + \int_V (e_{ijk}y_j E_{lk})_{,l} dV = 0.$$

That is,

$$\int_V e_{ijk}y_j(F_k + E_{lk,l})dV + \int_V e_{ijk}\delta_{jl} E_{lk} dV = 0.$$

The first integral is zero in virtue of (58.1) and we have

$$\int_V e_{ilk} E_{lk} dV = 0.$$

This equation is also an identity and consequently the integrand $e_{ilk}E_{lk}$ vanishes, from which we deduce that

$$E_{ij} = E_{ji}. \tag{58.2}$$

Thus the stress tensor is symmetric and the elastic equilibrium equations for a body are given by (58.1).

§ 59. Generalised Hooke's law

In the elementary theory of elasticity, Hooke's law states that the tension of an elastic string is proportional to the extension. In other words, stress is proportional to strain. The corresponding assumption in the general theory of elasticity is that the stress tensor is a linear homogeneous function of the strain tensor. That is

$$E_{ij} = c_{ijkl}e_{kl}. \tag{59.1}$$

It follows from the quotient law that c_{ijkl} is a Cartesian tensor of the fourth order, and it is called the **elasticity tensor**. Further from the symmetry of E_{ij} and e_{kl} we find that c_{ijkl} is symmetric not only with respect to the indices i and j but also with respect to k and l.

A body is said to be **homogeneous** if the elastic properties of the body are independent of the point under consideration. This means that the components of the elasticity tensor are all constants for a homogeneous body. We call a body **isotropic** if the elastic properties at a point are the same in all directions at that point. This means that the elasticity tensor c_{ijkl} transforms to c_{ijkl} itself under any rotation of axes.

§ 60. Isotropic tensors

A Cartesian tensor which transforms into itself under a rotation of axes is called an **isotropic tensor.** We have already met two isotropic tensors, namely δ_{ij} and e_{ijk}. We shall now search for the most general isotropic tensor c_{ijkl} of the fourth order. Its transformation law (55.1) becomes

$$c_{ijkl} = a_{ir}\, a_{js}\, a_{kt}\, a_{lu}\, c_{rstu}. \tag{60.1}$$

Let us rotate the axes through 180° about the y_3 axis. We deduce from (54.2) that $a_{ik} = -\delta_{ik} + 2l_i l_k$. But for this transformation $l_1 = l_2 = 0$ and $l_3 = 1$, and so the only non-zero components of a_{ik} are

$$a_{11} = -1, \quad a_{22} = -1, \quad a_{33} = +1.$$

By direct substitution in (60.1) we obtain $c_{ijkl} = -c_{ijkl}$, that is, $c_{ijkl} = 0$ in the following cases:-

(1) any three of the indices equal to 1 and the other equal to 3.

(2) any three of the indices equal to 2 and the other equal to 3.

(3) any two of the indices equal to 1, another equal to 2 and the other equal to 3.

(4) any two of the indices equal to 2, another equal to 1 and the other equal to 3.

Similar results are obtained by the corresponding rotations through 180° about the y_1 and the y_2 axes. Hence the only components which survive are those where the four indices are equal or equal in pairs.

Let us now rotate the axes through 90° about the y_3 axis. For such a transformation we deduce from (54.2) that $a_{ik} = l_i l_k + e_{ijk} l_j$ and so the only non-zero components of a_{ik} are

$$a_{12} = -1; \quad a_{21} = +1; \quad a_{33} = +1.$$

By direct substitution in (60.1) we discover that

$$\begin{gathered} c_{1111} = c_{2222}, \\ c_{1122} = c_{2211}, \; c_{1133} = c_{2233}, \; c_{3311} = c_{3322}, \\ c_{1212} = c_{2121}, \; c_{1313} = c_{2323}, \; c_{3131} = c_{3232}, \\ c_{1221} = c_{2112}, \; c_{1331} = c_{2332}, \; c_{3113} = c_{3223}. \end{gathered}$$

Similar results are obtained by the corresponding rotations through 90° about the y_1 and the y_2 axes. Hence we can collect all our results in the form

$$(60.2) \qquad \begin{cases} c_{iiii} = c_{jjjj}, \\ c_{iijj} = c_{iikk} = c_{llij} = c_{llkk}, \\ c_{ijij} = c_{ikik} = c_{ljlj} = c_{lklk}, \\ c_{ijji} = c_{ikki} = c_{ljjl} = c_{lkkl}, \end{cases}$$

where i, j, k and l are unequal and the summation convention does not apply. All other components are zero. The most general solution of equations (60.2) is then

$$(60.3) \quad c_{ijkl} = \lambda \delta_{ij} \delta_{kl} + \mu \delta_{ik} \delta_{jl} + \nu \delta_{il} \delta_{jk} + \kappa \delta_{ijkl},$$

where λ, μ, ν and κ are Cartesian invariants, and $\delta_{ijkl} = 1$ if all four indices are equal and otherwise zero.

Finally we carry out a *small* rotation, which is represented by $a_{ik} = \delta_{ik} + s_{ik}$, where s_{ik} is skew-symmetric and of the first order in small quantities. Substituting in (60.1) and retaining only terms of the first order we have

$$s_{ir} c_{rjkl} + s_{js} c_{iskl} + s_{kt} c_{ijtl} + s_{lu} c_{ijku} = 0.$$

Select $i = 2, j = k = l = 1$. Then in virtue of $s_{12} = -s_{21}$, the non-vanishing terms of this equation yield

$$c_{1111} = c_{2211} + c_{2121} + c_{2112}.$$

Substitute in this from (60.3) and the result is

$$\lambda + \mu + \nu + \kappa = \lambda + \mu + \nu$$

which gives $\kappa = 0$, so we can now write

$$c_{ijkl} = \lambda \delta_{ij} \delta_{kl} + \mu \delta_{ik} \delta_{jl} + \nu \delta_{il} \delta_{jk}. \tag{60.4}$$

This tensor is obviously isotropic and so it is the most general isotropic tensor of the fourth order.

Ex. 1. Prove that the most general isotropic tensors of the second and third orders are $\lambda \delta_{ij}$ and λe_{ijk} respectively where λ is an invariant.

Ex. 2. Prove that c_{ijkl} defined by (60.3) satisfies the symmetry relations $c_{ijkl} = c_{klij}$ and $c_{ijkl} = c_{jilk}$.

§ 61. Homogeneous and isotropic body

In this section we confine our attention to a body which is both homogeneous and isotropic. Then the elasticity tensor is isotropic with constant components. We therefore require an isotropic tensor c_{ijkl} which is symmetric in both i and j and in k and l, and whose components are all constants. To obtain this tensor, substitute from (60.4) into the equation $c_{ijkl} = c_{ijlk}$ and we obtain

$$(\mu - \nu)(\delta_{ik} \delta_{jl} - \delta_{il} \delta_{jk}) = 0.$$

In this equation put $i = k = 1, j = l = 2$ and we see that $\mu = \nu$. Thus the isotropic tensor

$$c_{ijkl} = \lambda \delta_{ij} \delta_{kl} + \mu(\delta_{ik} \delta_{jl} + \delta_{il} \delta_{jk}) \tag{61.1}$$

satisfies the required symmetry. The generalised Hooke's law (59.1) for a homogeneous, isotropic body now becomes

$$E_{ij} = [\lambda\delta_{ij}\delta_{kl} + \mu(\delta_{ik}\delta_{jl} + \delta_{il}\delta_{jk})]e_{kl}$$

where λ and μ are *constants*. This equation simplifies to

$$E_{ij} = \lambda\theta\delta_{ij} + 2\mu e_{ij}, \tag{61.2}$$

where θ is the dilatation defined by (56.5). Contracting i and j, we obtain the equation

$$\Theta = E_{ii} = (3\lambda + 2\mu)\theta \tag{61.3}$$

which connects the two invariants θ and Θ.

We now solve equations (61.2) for the strain tensor in terms of the stress tensor and in virtue of (61.3) we obtain

$$e_{ij} = -\frac{\lambda\Theta}{2\mu(3\lambda + 2\mu)}\delta_{ij} + \frac{1}{2\mu}E_{ij}. \tag{61.4}$$

It is usual to work in terms of **Young's modulus** E and **Poisson's ratio** σ defined by

$$E = \frac{\mu(3\lambda + 2\mu)}{\lambda + \mu}; \quad \sigma = \frac{\lambda}{2(\lambda + \mu)}. \tag{61.5}$$

Equation (61.4) then becomes

$$e_{ij} = \frac{1}{E}\{-\sigma\Theta\delta_{ij} + (1 + \sigma)E_{ij}\}. \tag{61.6}$$

We can obtain stress compatibility equations by substituting this expression for e_{ij} in (56.6).

Sometimes we require the equations of equilibrium in terms of the displacement vector u_i. To obtain these we substitute from (61.2) in (58.1) and obtain

$$F_j + \lambda\theta_{,i}\delta_{ij} + 2\mu e_{ij,i} = 0$$

which becomes by (56.2)

$$F_j + \lambda\theta_{,j} + \mu(u_{i,ji} + u_{j,ii}) = 0.$$

But $\theta = e_{ii} = \frac{1}{2}(u_{i,i} + u_{i,i}) = u_{i,i}$. Consequently $\theta_{,j} = u_{i,ij}$, and the above equations take the form

$$F_j + (\lambda + \mu)\theta_{,j} + \mu\nabla^2 u_j = 0 \tag{61.7}$$

where ∇^2 is the Laplacian operator. Now that we have reverted to displacements u_i we do not require any compatibility equations.

Ex. Show that if an isotropic, homogeneous body is in equilibrium under no forces, then $\nabla^2\theta = 0$.

§ 62. Curvilinear coordinates

Many problems in elasticity can be investigated more conveniently by means of curvilinear coordinates x^i in which the line-element ds is given by $ds^2 = g_{ij}\,dx^i dx^j$. In this case we must return to the distinction between contravariance and covariance.

The coordinate system serves only to describe the actual strains and stresses. The laws of elasticity are themselves independent of the coordinates. Therefore these laws can be formulated as tensor equations. We recall (section 9) that if a tensor is zero in *any one* coordinate system, it is zero in *all* coordinate systems. Consequently if we write down tensor equations, which reduce in cartesian coordinates to the results already established, then these tensor equations express the theory with respect to any curvilinear system. Accordingly we immediately verify that the laws of elasticity are represented by the following tensor equations:-

$$e = e_{ij}l^i l^j, \; e_{ij} = \tfrac{1}{2}(u_{i,j} + u_{j,i}), \; \theta = g^{ij}e_{ij},$$
$$e_{ij,kl} + e_{kl,ij} - e_{ik,jl} - e_{jl,ik} = 0,$$
$$T_i = E_{ji}n^j, \; E_{ij} = E_{ji}, \; F^j + E^{ij}_{,i} = 0,$$
$$E_{ij} = c_{ijkl}e^{kl}, \; c_{ijkl} = c_{jikl} = c_{ijlk}\,*.$$

* It can also be proved that $c_{klij} = c_{ijkl}$. See I. S. Sokolnikoff, Mathematical Theory of Elasticity, section 26.

If the body is homogeneous and isotropic, we have in addition

$$c_{ijkl} = \lambda g_{ij} g_{kl} + \mu(g_{ki} g_{lj} + g_{kj} g_{li}),$$

$$E_{ij} = \lambda\theta g_{ij} + 2\mu e_{ij}, \quad \Theta = g^{ij} E_{ij} = (3\lambda + 2\mu)\theta,$$

$$e_{ij} = \frac{1}{E}\{-\sigma\Theta g_{ij} + (1+\sigma)E_{ij}\}$$

$$F_j + (\lambda + \mu)\theta_{,j} + \mu g^{rs} u_{j,rs} = 0.$$

In all these equations l^i is a *unit* vector specifying the direction of the extension e, n^i is the *unit* vector normal to the small element ΔS, and commas once more denote covariant differentiation.

It is important to note that the components of the vectors u_i, T_i and F_i may not possess a *physical* significance of the correct dimensionality. For example, the exercise of section 5 shows us that the second component of the acceleration vector in polar coordinates is an *angular* acceleration. It will suffice to discuss the force vector F_i, whose components in a cartesian system are $\bar{F}_i = \bar{F}^i$ (say). Then $F_i = \frac{\partial \bar{x}^j}{\partial x^i} \bar{F}_j$ and $F^i = \frac{\partial x^i}{\partial \bar{x}^j} \bar{F}^j$ are the covariant and contravariant components in the x^i system, where we have put $\bar{x}^i = y_i$ the cartesian variables. The component of $\bar{F}^i$, which is the physical force vector, at any point in the direction of the unit vector $\bar{l}^i = \bar{l}_i$ is the invariant $\bar{F}^i \bar{l}_i = F^i l_i = g_{ij} F^i l^j$. The contravariant unit vector in the direction of the x^1 axis is $\delta^i_1/\sqrt{g_{11}}$. Therefore the actual physical components of force along the x^1 coordinate curve are $g_{ij} F^i \delta^j_1/\sqrt{g_{11}} = g_{i1} F^i/\sqrt{g_{11}}$. In the exercise of section 5, we have $g_{11} = 1$, $g_{12} = 0$, $g_{22} = r^2$ and so the physical components of acceleration in polar coordinates are $\frac{d^2 r}{dt^2} - r\left(\frac{d\theta}{dt}\right)^2$ and $r\frac{d^2\theta}{dt^2} + 2\frac{dr}{dt}\frac{d\theta}{dt}$, which represent the radial and transverse accelerations respectively.

Similarly the tensors e_{ij} and E_{ij} have no direct physical significance. Consider the strain tensor e_{ij} which is connected with the cartesian strain tensor $\bar{e}_{kl}$ by $e_{ij} = \dfrac{\partial \bar{x}^k}{\partial x^i}\dfrac{\partial \bar{x}^l}{\partial x^j}\bar{e}_{kl}$. The physical component of the strain tensor associated with the directions of the unit vectors l^i and m^i at a point may be defined to be the invariant $e_{ij}l^i m^j$.

As an example, let us discuss cylindrical coordinates r, θ and z given by

$$y_1 = r\cos\theta, \quad y_2 = r\sin\theta, \quad y_3 = z.$$

The non-zero components of the fundamental tensor are

$$g_{11} = 1, \quad g_{22} = r^2, \quad g_{33} = 1$$

from which we deduce that

$$g^{11} = 1, \quad g^{22} = \frac{1}{r^2}, \quad g^{33} = 1, \quad g^{ij} = 0 \text{ if } i \neq j.$$

So the only surviving Christoffel symbols of the second kind are

$$\begin{Bmatrix} 1 \\ 22 \end{Bmatrix} = -r, \quad \begin{Bmatrix} 2 \\ 12 \end{Bmatrix} = \frac{1}{r}.$$

The actual physical components $(u_\alpha, u_\beta, u_\gamma)$ of the displacement vector u_i are

$$u_\alpha = u_1, \quad u_\beta = \frac{1}{r}u_2, \quad u_\gamma = u_3.$$

Further calculation will show that the components of the strain tensor are

$$e_{11} = \frac{\partial u_1}{\partial r}, \quad e_{22} = \frac{\partial u_2}{\partial \theta} + ru, \quad e_{33} = \frac{\partial u_3}{\partial z},$$

$$e_{12} = \tfrac{1}{2}\left(\frac{\partial u_1}{\partial \theta} + \frac{\partial u_2}{\partial r} - \frac{2u_2}{r}\right), \quad e_{23} = \tfrac{1}{2}\left(\frac{\partial u_2}{\partial z} + \frac{\partial u_3}{\partial \theta}\right),$$

$$e_{31} = \tfrac{1}{2}\left(\frac{\partial u_3}{\partial r} + \frac{\partial u_1}{\partial z}\right),$$

whilst the physical components associated with the directions of the coordinate curves are

$$e_{\alpha\alpha} = \frac{\partial u_\alpha}{\partial r}, \quad e_{\beta\beta} = \frac{1}{r}\frac{\partial u_\beta}{\partial \theta} + \frac{u_\alpha}{r}, \quad e_{\gamma\gamma} = \frac{\partial u_\gamma}{\partial z},$$

$$e_{\alpha\beta} = \tfrac{1}{2}\left(\frac{1}{r}\frac{\partial u_\alpha}{\partial \theta} + \frac{\partial u_\beta}{\partial r} - \frac{u_\beta}{r}\right), \quad e_{\alpha\gamma} = \tfrac{1}{2}\left(\frac{\partial u_\gamma}{\partial r} + \frac{\partial u_\alpha}{\partial z}\right),$$

$$e_{\gamma\beta} = \tfrac{1}{2}\left(\frac{\partial u_\beta}{\partial z} + \frac{1}{r}\frac{\partial u_\gamma}{\partial \theta}\right).$$

The dilatation is

$$\theta = g^{ij} e_{ij} = \frac{\partial u_\alpha}{\partial r} + \frac{1}{r}\frac{\partial u_\beta}{\partial \theta} + \frac{\partial u_\gamma}{\partial z} + \frac{u_\alpha}{r}.$$

Ex. Find the physical components associated with the directions of the coordinate curves of the strain tensor in terms of the physical components of the displacement vector when the coordinates are spherical polars.

§ 63. Mechanics of continuous matter

Let us discuss the motion of a continuous medium by the Eulerian method. Instead of following the path traced out by a particular particle, we focus our attention on a definite point P of the medium whose coordinates referred to a cartesian coordinate system are y_i. Let us denote by u_i the velocity vector of that particle which happens to be at P at time t. Then u_i is a function of y_i and t. After a further interval of time Δt has elapsed, the particle which was originally at P is now at the point $y_i + u_i \Delta t$ with velocity $u_i + \Delta u_i$. Hence $u_i + \Delta u_i$ is the function u_i at the point $y_i + u_i \Delta t$ and at time $t + \Delta t$. On expanding by Taylor's theorem we have

$$u_i + \Delta u_i = u_i + u_j \Delta t u_{i,j} + \Delta t \frac{\partial u_i}{\partial t}.$$

The acceleration vector f_i at P is the limit of $\Delta u_i/\Delta t$ as Δt tends to zero. We therefore deduce that the acceleration vector is given by

$$(63.1) \qquad f_i = u_j u_{i,j} + \frac{\partial u_i}{\partial t}.$$

Now consider a volume V of the medium, bounded by the surface S. The mass M contained inside V is $M = \int_V \rho dV$, where the density ρ is a function of y_i and t. Thus the rate of increase of mass is $\frac{dM}{dt} = \int_V \frac{\partial \rho}{\partial t} dV$. Let n_i denote the direction cosines of the external normal to the small element of surface ΔS. Then the rate of mass flow outwards across that element ΔS is $\rho n_i u_i \Delta S$. Hence the rate of increase of mass is also expressed by the surface integral $-\int_S \rho n_i u_i dS$. Consequently we have

$$\int_S \rho n_i u_i dS + \int_V \frac{\partial \rho}{\partial t} dV = 0.$$

We apply Gauss's theorem to the surface integral and the result is

$$\int_V (\rho u_i)_{,i} dV + \int_V \frac{\partial p}{\partial t} dV = 0.$$

This equation is an identity and so we derive the **equation of continuity**

$$(63.2) \qquad (\rho u_i)_{,i} + \frac{\partial \rho}{\partial t} = 0.$$

Physically this equation expresses the principle of conservation of mass.

In section 58 we discussed the equilibrium of a continuous medium. The equations of motion can be determined in the same way. If we replace F_j by $F_j - \rho f_j$ these equations become

$$(63.3) \qquad \rho f_j = F_j + E_{ij,i}.$$

We now substitute for f_j from (63.1) and obtain

$$\rho u_i u_{j,i} + \rho \frac{\partial u_j}{\partial t} = F_j + E_{ij,i}.$$

This equation can be written, in virtue of (63.2), in the form

(63.4) $$[\rho u_i u_j - E_{ij}]_{,i} + \frac{\partial}{\partial t}(\rho u_j) = F_j.$$

(63.2) and (63.4) constitute the equations of motion of a continuous medium.

When the coordinate system is curvilinear the equations (63.2) and (63.4) may be expressed in the tensor forms

(63.5) $$(\rho u^i)_{,i} + \frac{\partial \rho}{\partial t} = 0,$$

(63.6) $$(\rho u^i u^j - E^{ij})_{,i} + \frac{\partial}{\partial t}(\rho u^j) = F^j.$$

Solutions

p. 88. Ex. The only non-vanishing components of a_{ik} are

$$a_{12} = -1; \; a_{21} = +1; \; a_{33} = 1.$$

p. 103. Ex. $$e_{\alpha\alpha} = \frac{\partial u_\alpha}{\partial r}; \quad e_{\beta\beta} = \frac{1}{r}\frac{\partial u_\beta}{\partial \theta} + \frac{u_\alpha}{r};$$

$$e_{\gamma\gamma} = \frac{1}{r \sin\theta}\frac{\partial u_\gamma}{\partial \psi} + \frac{u_\alpha}{r} + \frac{\cot\theta}{r} u_\beta;$$

$$e_{\beta\gamma} = \frac{1}{2}\left(\frac{1}{r\sin\theta}\frac{\partial u_\beta}{\partial \psi} + \frac{1}{r}\frac{\partial u_\gamma}{\partial \theta} - \frac{\cot\theta}{r} u_\gamma\right);$$

$$e_{\gamma\alpha} = \frac{1}{2}\left(\frac{1}{r\sin\theta}\frac{\partial u_\alpha}{\partial \psi} + \frac{\partial u_\gamma}{\partial r} - \frac{u_\gamma}{r}\right);$$

$$e_{\alpha\beta} = \frac{1}{2}\left(\frac{1}{r}\frac{\partial u_\alpha}{\partial \theta} + \frac{\partial u_\beta}{\partial r} - \frac{u_\beta}{r}\right).$$

CHAPTER VIII

THEORY OF RELATIVITY

§ 64. Special theory

In classical mechanics, the position of a point in space at which an event occurs can be determined by its three space coordinates x^1, x^2, x^3 referred to some rectangular cartesian system. Also an observer can measure the time t at which the event takes place by means of a clock. An event is then fixed in both space and time by the **system** S which is comprised of the four numbers x^1, x^2, x^3 and t.

Einstein examined the concept 'simultaneity' and came to the conclusion that 'simultaneous events at different points' has no meaning without further qualification. Continuing his study of fundamental ideas, Einstein arrived at the Special Theory of Relativity which he based on the two principles: (1) *it is impossible to detect the unaccelerated translatory motion of a system through space,* (2) *the velocity c of a ray of light is a constant which does not depend on the relative velocity of its source and the observer.*

Let us now consider two systems S and $\bar{S}$ which coincide at the time $t = 0$, such that $\bar{S}$ moves with constant velocity V along the x^1 axis of the S system. Then the **Lorentz Transformation**, which can be deduced from the two principles of special relativity, connects the space coordinates and the time of both systems by the equations

$$\bar{x}^1 = \beta(x^1 - Vt), \quad \bar{x}^2 = x^2, \quad \bar{x}^3 = x^3, \quad \bar{t} = \beta(t - Vx^1/c^2),$$

where $\beta = (1 - V^2/c^2)^{-1/2}$. We easily verify that

$$\begin{aligned} &- (d\bar{x}^1)^2 - (d\bar{x}^2)^2 - (d\bar{x}^3)^2 + c^2(d\bar{t})^2 \\ &= - (dx^1)^2 - (dx^2)^2 - (dx^3)^2 + c^2(dt)^2. \end{aligned}$$

The invariance of this equation with respect to Lorentz transformations suggests that the **Minkowski space** defined by the metric

$$(64.1) \quad d\sigma^2 = -(dx^1)^2 - (dx^2)^2 - (dx^3)^2 + c^2(dx^4)^2$$

where we have written $x^4 = t$, is appropriate for the geometrical discussion of special relativity. We denote the line-element of this four-dimensional space by $d\sigma$ (not by ds) in order to emphasise that $d\sigma$ is not the physical distance between two neighbouring points.

The Minkowski space is flat and its signature, which equals the excess of the number of positive terms over the number of negative terms in its metric, is -2. It is well-known* that there is no *real* transformation of coordinates which will reduce (64.1) to the metric of a four-dimensional Euclidean space, whose signature is 4. Thus the geometry of Minkowski space differs in many respects from Euclidean geometry; for example, there exist real null curves (see (15.2)) and real null-geodesics.

In this chapter Latin indices will have the range 1 to 3 whilst Greek indices will range from 1 to 4. The velocity u of a particle, which is at the point x^i has the components $u^i = dx^i/dt$ referred to the system S. It follows from (64.1) that

$$(64.2) \qquad \frac{dt}{d\sigma} = \frac{1}{c}\left(1 - \frac{u^2}{c^2}\right)^{-1/2}.$$

The four-dimensional Minkowski **momentum vector** is defined by $m_0 c \dfrac{dx^\alpha}{d\sigma}$, where m_0 is a constant. The special theory identifies the fourth component $m_0 c \dfrac{dx^4}{d\sigma}$ with the mass m of the moving particle. In virtue of (64.2) we have

$$(64.3) \qquad m = m_0 \left(1 - \frac{u^2}{c^2}\right)^{-1/2}.$$

* W. L. Ferrar, Algebra, p. 154.

The constant m_0 is the mass when $u = 0$ and so it is called the **rest-mass** of the particle. The mass m, which clearly increases with the velocity, is called the **relativistic mass** of the particle. The components

$$m_0 c \frac{dx^i}{d\sigma} = m_0 c \frac{dx^i}{dt} \frac{dt}{d\sigma} = m \frac{dx^i}{dt}$$

and are evident generalizations of the Newtonian momentum vector.

We define the four-dimensional Minkowski **force vector** F^α by

(64.4)

$$F^\alpha = m_0 c^2 \frac{d^2 x^\alpha}{d\sigma^2} = c^2 \frac{d}{d\sigma}\left(m_0 \frac{dx^\alpha}{d\sigma}\right) = \left(1 - \frac{u^2}{c^2}\right)^{-1/2} \frac{d}{dt}\left(m \frac{dx^\alpha}{dt}\right).$$

The Newtonian force vector is $X^i = \dfrac{d}{dt}\left(m \dfrac{dx^i}{dt}\right)$ and so

$$F^i = \left(1 - \frac{u^2}{c^2}\right)^{-1/2} X^i.$$

We obtain by expansion from (64.3) that

$$mc^2 = m_0 c^2 + \frac{1}{2} m_0 u^2 + \ldots$$

and so the special theory identifies the energy E associated with a particle by the equation $E = mc^2$. Therefore

$$F^4 = \left(1 - \frac{u^2}{c^2}\right)^{-1/2} \frac{dm}{dt} = \frac{1}{c^2}\left(1 - \frac{u^2}{c^2}\right)^{-1/2} \frac{dE}{dt}.$$

The motion of a particle which moves under the action of some force system can be represented in Minkowski space by a curve, called the **world-line** of the particle. If no forces act on the particle, we see from (64.4) that $d^2 x^\alpha / d\sigma^2 = 0$. Thus the world-line of a free particle is a geodesic of the Minkowski space.

The velocity of a light ray is the constant c, and so

we see from (64.1) that for such a ray $d\sigma = 0$. Accordingly the world-line of a light ray is a null-geodesic of the Minkowski space.

In order to discuss the mechanics of a continuous medium, we introduce the symmetrical four-dimensional **energy-momentum tensor** $T^{\alpha\beta}$ defined by

$$T^{ij} = T^{ji} = \rho u^i u^j - E^{ij};\ T^{i4} = T^{4i} = \rho u^i;\ T^{44} = \rho,$$

where ρ is the density and E^{ij} is the Cartesian stress tensor defined in section 57. Then the special theory generalises (63.5) and (63.6), which are the equations of motion of a continuous medium into

$$T^{\alpha\beta}_{,\alpha} = F^{\beta}. \tag{64.5}$$

If we change to spherical polar coordinates r, θ and ψ, the metric of Minkowski space becomes

$$d\sigma^2 = -dr^2 - r^2 d\theta^2 - r^2 \sin^2 \theta d\psi^2 + c^2 dt^2. \tag{64.6}$$

§ 65. Maxwell's Equations

The classical theory of electrodynamics*, according to Lorentz, is specified by the **electric potential** φ which is a scalar and the **magnetic potential** A_i which is a vector. The **electric field strength** vector E_i and the **magnetic field strength** vector H_i are derived from these potentials by the equations

$$E_i = -\operatorname{grad} \varphi - \frac{1}{c}\frac{\partial A_i}{\partial t},$$

$$H_i = \operatorname{curl} A_i.$$

* M. Abraham—R. Becker, Electricity and Magnetism.
C. A. Coulson, Waves, chap. VII.
D. E. Rutherford, Vector Methods, p. 126.

Using electrostatic units, Maxwell's equations are

$$
(65.1) \qquad \begin{cases} \operatorname{div} E_i = 4\pi\rho, \\ \operatorname{div} H_i = 0, \\ \operatorname{curl} E_i + \dfrac{1}{c}\dfrac{\partial H_i}{\partial t} = 0, \\ \operatorname{curl} H_i - \dfrac{1}{c}\dfrac{\partial E_i}{\partial t} = \dfrac{4\pi}{c} j_i, \end{cases}
$$

where j_i is the **current density** vector and ρ is the **charge density.**

In Minkowski space, with the metric (64.1), let us form the four-dimensional potential vector Φ_α and the four-dimensional current density vector J^α defined respectively by

$$
\Phi_\alpha \equiv (-A_1, -A_2, -A_3, c\varphi),
$$
$$
J_\alpha \equiv (j_1, j_2, j_3, \rho),
$$

with respect to a particular coordinate system. Next we introduce the skew-symmetrical tensor $\eta_{\alpha\beta}$ defined by

$$
\eta_{\alpha\beta} = \Phi_{\alpha,\beta} - \Phi_{\beta,\alpha} = \frac{\partial \Phi_\alpha}{\partial x^\beta} - \frac{\partial \Phi_\beta}{\partial x^\alpha}
$$

and we immediately calculate that its non-vanishing components in the given coordinate system are

$$
\eta_{23} = -\eta_{32} = H_1;\ \eta_{31} = -\eta_{13} = H_2;\ \eta_{12} = -\eta_{21} = H_3;
$$
$$
\eta_{14} = -\eta_{41} = cE_1;\ \eta_{24} = -\eta_{42} = cE_2;\ \eta_{34} = -\eta_{43} = cE_3.
$$

The non-vanishing contravariant components $\eta^{\alpha\beta}$ may now be obtained and are

$$
\eta^{23} = -\eta^{32} = H_1;\quad \eta^{31} = -\eta^{13} = H_2;\quad \eta^{12} = -\eta^{21} = H_3;
$$
$$
\eta^{14} = -\eta^{41} = -\frac{E_1}{c};\ \eta^{24} = -\eta^{42} = -\frac{E_2}{c};\ \eta^{34} = -\eta^{43} = -\frac{E_3}{c}.
$$

We now write Maxwell's equations (65.1) in terms of

η and J and the results are readily verified to be respectively

$$\frac{\partial \eta^{41}}{\partial x^1} + \frac{\partial \eta^{42}}{\partial x^2} + \frac{\partial \eta^{43}}{\partial x^3} = \frac{4\pi}{c} J^4,$$

$$\frac{\partial \eta_{23}}{\partial x^1} + \frac{\partial \eta_{31}}{\partial x^2} + \frac{\partial \eta_{12}}{\partial x^3} = 0,$$

$$\frac{\partial \eta_{ij}}{\partial x^4} + \frac{\partial \eta_{j4}}{\partial x^i} + \frac{\partial \eta_{4i}}{\partial x^j} = 0,$$

$$\frac{\partial \eta^{i1}}{\partial x^1} + \frac{\partial \eta^{i2}}{\partial x^2} + \frac{\partial \eta^{i3}}{\partial x^3} + \frac{\partial \eta^{i4}}{\partial x^4} = \frac{4\pi}{c} J^i.$$

The first and last of these equations combine together into the form

$$\eta^{\alpha\beta}_{,\beta} = \frac{4\pi}{c} J^\alpha, \tag{65.2}$$

whilst the remaining two are accounted for by the equations of the set

$$\eta_{\alpha\beta,\gamma} + \eta_{\beta\gamma,\alpha} + \eta_{\gamma\alpha,\beta} = 0 \tag{65.3}$$

which do not vanish identically.

We have accordingly written Maxwell's equations in tensor form in Minkowski space. Thus they are invariant under the Lorentz group of transformations.

§ 66. General theory

We now turn to the General Theory of Relativity which was developed by Einstein in order to discuss gravitation. He postulated the **principle of covariance**, which asserts that the laws of physics must be independent of the space-time coordinates. This swept away the privileged role of the Lorentz transformation. As a result Minkowski space was replaced by the Riemannian V_4 with the general metric

$$d\sigma^2 = g_{\alpha\beta}\, dx^\alpha\, dx^\beta. \tag{66.1}$$

Einstein also introduced the **principle of equivalence,**

which in essence states that the fundamental tensor $g_{\alpha\beta}$ can be chosen to account for the presence of a gravitational field. That is, $g_{\alpha\beta}$ depends on the distribution of matter and energy in physical space.

Matter and energy can be specified by the energy-momentum tensor $T^{\alpha\beta}$ which in the special theory satisfies the equation $T^{\alpha\beta}_{\ ,\alpha} = F^{\beta}$. The only forces, namely those due to gravitation, are however already taken into account by the choice of the fundamental tensor $g_{\alpha\beta}$. We therefore ignore F^{β} and, in accordance with the principle of covariance, the energy-momentum tensor must now satisfy the equation $T^{\alpha\beta}_{\ ,\alpha} = 0$. We shall write this equation in the equivalent form $T^{\alpha}_{.\beta,\alpha} = 0$ where $T^{\alpha}_{.\beta} = g_{\beta\gamma}T^{\alpha\gamma}$ is the mixed energy-momentum tensor. The problem now is to determine $T^{\alpha}_{.\beta}$ as a function of the $g_{\alpha\beta}$ and their derivatives up to the second order, bearing in mind that $T^{\alpha}_{.\beta,\alpha} = 0$. We recall from (34.3) that Einstein's tensor defined by

$$G^{\alpha}_{.\beta} = g^{\alpha\gamma} R_{\beta\gamma} - \tfrac{1}{2}R\delta^{\alpha}_{\beta} \tag{66.2}$$

satisfies the equation $G^{\alpha}_{.\beta,\alpha} = 0$. The equations of motion require $T^{\alpha}_{.\beta,\alpha} = 0$, but very remarkably $G^{\alpha}_{.\beta,\alpha} = 0$ is an identity in Riemannian geometry. This led Einstein to propose the relation

$$\kappa T^{\alpha}_{.\beta} + G^{\alpha}_{.\beta} = 0. \tag{66.3}$$

In effect these equations form the link between the physical energy-momentum tensor $T^{\alpha}_{.\beta}$ and the geometrical tensor $G^{\alpha}_{.\beta}$ of the V_4 of general relativity. In order that Newton's theory of gravitation can be deduced as a first approximation from Einstein's theory, it was found necessary to choose $\kappa = 8\pi k/c^4$ where k is the usual gravitational constant 6.664×10^{-8} cm^3. gm.$^{-1}$ sec.$^{-2}$. The value of c is 2.99796×10^{10} cm. sec.$^{-1}$, and so κ is 2.073×10^{-48} cm.$^{-1}$ gm.$^{-1}$ sec.^{2} in c.g.s. units.

In the special theory, the world-lines of free particles

and of light rays are respectively the geodesics and the null-geodesics of Minkowski space. The principle of equivalence demands that all particles be regarded as free particles when gravitation is the only force under consideration. Then it follows from the principle of covariance that the world-line of a particle under the action of gravitational forces is a geodesic of the V_4 with the metric (66.1). Similarly the world-line of a light ray is a null-geodesic.

§ 67. Spherically symmetrical metric

General relativity discusses several important problems in which the coordinate system r, θ, ψ and t is such that the metric takes the form

$$(67.1)\quad d\sigma^2 = -e^\lambda dr^2 - r^2 d\theta^2 - r^2 \sin^2\theta\, d\psi^2 + c^2 e^\nu dt^2,$$

where λ and ν are functions of r. A metric of this type is said to be **spherically symmetrical.** It is a generalisation of the special relativity metric (64.6), which is expressed in spherical polars. The coefficients of dr^2 and dt^2 have been selected as exponentials in order to ensure that the signature of $d\sigma^2$ is -2. Let us write $x^1 = r$, $x^2 = \theta$, $x^3 = \psi$ and $x^4 = ct$. Then the non-zero components of the fundamental tensor are

$$g_{11} = -e^\lambda,\ g_{22} = -r^2,\ g_{33} = -r^2\sin^2\theta,\ g_{44} = e^\nu.$$

The determinant g becomes

$$g = -r^4 \sin^2\theta\, e^{\lambda+\nu}$$

and hence the non-zero components of the conjugate symmetric fundamental tensor are

$$g^{11} = -e^{-\lambda},\ g^{22} = -\frac{1}{r^2},\ g^{33} = -\frac{1}{r^2\sin^2\theta},\ g^{44} = e^{-\nu}.$$

A brief calculation shows that the only non-vanishing Christoffel symbols of the second kind are

$$(67.2)\begin{cases} \begin{Bmatrix}1\\11\end{Bmatrix} = \frac{1}{2}\lambda', & \begin{Bmatrix}2\\12\end{Bmatrix} = \frac{1}{r}, \qquad \begin{Bmatrix}3\\31\end{Bmatrix} = \frac{1}{r}, \\ \begin{Bmatrix}1\\22\end{Bmatrix} = -re^{-\lambda}, & \begin{Bmatrix}2\\33\end{Bmatrix} = -\sin\theta\cos\theta, \quad \begin{Bmatrix}3\\23\end{Bmatrix} = \cot\theta, \\ \begin{Bmatrix}1\\33\end{Bmatrix} = -r\sin^2\theta\; e^{-\lambda}, & \\ & \begin{Bmatrix}4\\14\end{Bmatrix} = \frac{1}{2}\nu', \\ \begin{Bmatrix}1\\44\end{Bmatrix} = \frac{1}{2}\nu' e^{-\lambda+\nu}, & \end{cases}$$

where a dash denotes differentiation with respect to r. We now evaluate the components of the Ricci tensor by means of (33.2) and the results are

$$(67.3)\begin{cases} R_{11} = -\frac{1}{r}\lambda' - \frac{1}{4}\lambda'\nu' + \frac{1}{2}\nu'' + \frac{1}{4}\nu'^2, \\ R_{22} = \operatorname{cosec}^2\theta\, R_{33} = -1 + e^{-\lambda}\{1 - \frac{1}{2}r\lambda' + \frac{1}{2}r\nu'\}, \\ R_{44} = e^{-\lambda+\nu}\{\frac{1}{4}\lambda'\nu' - \frac{1}{2}\nu'' - \frac{1}{r}\nu' - \frac{1}{4}\nu'^2\}, \\ R_{\alpha\beta} = 0 \text{ for } \alpha \neq \beta. \end{cases}$$

A further calculation gives us the curvature invariant

$$(67.4)\quad R = \frac{2}{r^2} + e^{-\lambda}\left\{-\frac{2}{r^2} + \frac{2}{r}\lambda' + \frac{1}{2}\lambda'\nu' - \nu'' - \frac{2}{r}\nu' - \frac{1}{2}\nu'^2\right\}.$$

On substituting this in (66.2) we obtain the components of the Einstein tensor for the spherically symmetrical metric (67.1)

$$(67.5)\begin{cases} G^1_{.1} = -\frac{1}{r^2} + e^{-\lambda}\left\{\frac{1}{r^2} + \frac{1}{r}\nu'\right\}, \\ G^2_{.2} = G^3_{.3} = e^{-\lambda}\left\{-\frac{1}{2r}\lambda' - \frac{1}{4}\lambda'\nu' + \frac{1}{2}\nu'' + \frac{1}{2r}\nu' + \frac{1}{4}\nu'^2\right\}. \\ G^4_{.4} = -\frac{1}{r^2} + e^{-\lambda}\left\{\frac{1}{r^2} - \frac{1}{r}\lambda'\right\}, \\ G^\alpha_{.\beta} = 0 \text{ for } \alpha \neq \beta. \end{cases}$$

Ex. Find the necessary and sufficient conditions that a space with a spherically symmetrical metric be an Einstein space.

§ 68. Schwarzschild metric

We now seek the spherically symmetrical metric (67.1) consistent with the existence of one gravitating point particle situated at the origin, and surrounded by empty space. When the origin itself is excluded from our discussion, the energy-momentum tensor $T^{\alpha}_{\cdot\beta}$ is zero at all points. It follows from (66.3) that $G^{\alpha}_{\cdot\beta} = 0$, which yields in virtue of (67.5) the equations

$$(68.1) \quad -\frac{1}{r^2} + e^{-\lambda}\left\{\frac{1}{r^2} + \frac{1}{r}\nu'\right\} = 0,$$

$$(68.2) \quad e^{-\lambda}\left\{-\frac{1}{2r}\lambda' - \frac{1}{4}\lambda'\nu' + \frac{1}{2}\nu'' + \frac{1}{2r}\nu' + \frac{1}{4}\nu'^2\right\} = 0,$$

$$(68.3) \quad -\frac{1}{r^2} + e^{-\lambda}\left\{\frac{1}{r^2} - \frac{1}{r}\lambda'\right\} = 0.$$

The solution of (68.3) is readily found to be $e^{-\lambda} = 1 - 2m/(c^2r)$, where the constant of integration m introduced in this way can be identified physically with the rest mass of the gravitating particle. Further, we subtract (68.3) from (68.1) and obtain $e^{-\lambda}(\lambda' + \nu')/r = 0$. That is $\lambda' + \nu' = 0$, from which we have $\lambda + \nu = k$, where k is a constant. Thus $e^{\nu} = e^{k}\{1 - 2m/(c^2r)\}$. We can now verify that equation (68.2) is satisfied identically. However, at large distances from the gravitating point, the metric should approximate to the metric (64.6) of special relativity. Therefore, we must select $k = 0$. We thus obtain the **Schwarzschild metric**

$$(68.4) \quad d\sigma^2 = -\left(1 - \frac{2m}{c^2 r}\right)^{-1} dr^2 - r^2\, d\theta^2 - r^2 \sin^2\theta\, d\varphi^2 + c^2\left(1 - \frac{2m}{c^2 r}\right) dt^2.$$

Ex. Show that a space with Schwarzschild's metric is an Einstein space, but not a space of constant curvature.

§ 69. Planetary motion

Let us investigate the motion of a planet in the gravitational field of the sun. The sun will be selected as a gravitating particle and the planet as a free particle whose mass is so small that it does not affect the metric, and whose world-line is then a geodesic in the V_4 with the Schwarzschild metric (68.4). The geodesics are determined by the four equations (26.4) in which we now, of course, replace s by σ. We shall omit one of these equations, in practice the most formidable one involving $d^2r/d\sigma^2$, and replace it by (27.1) which is of the first order and is satisfied along the geodesics. When this is done, we no longer require the Christoffel symbols of the type $\left\{ {1 \atop \alpha\beta} \right\}$. The remaining non-vanishing symbols of the second kind can be calculated from (67.2) and may be found to be

$$\left\{ {2 \atop 12} \right\} = \frac{1}{r}, \qquad \left\{ {3 \atop 13} \right\} = \frac{1}{r}, \qquad \left\{ {4 \atop 14} \right\} = \frac{m}{c^2 r^2}\left(1 - \frac{2m}{c^2 r}\right)^{-1}.$$

$$\left\{ {2 \atop 33} \right\} = -\sin\theta\cos\theta, \quad \left\{ {3 \atop 23} \right\} = \cot\theta,$$

Hence the four equations of the geodesics are

$$\frac{d^2\theta}{d\sigma^2} + \frac{2}{r}\frac{dr}{d\sigma}\frac{d\theta}{d\sigma} - \sin\theta\cos\theta\left(\frac{d\psi}{d\sigma}\right)^2 = 0, \tag{69.1}$$

$$\frac{d^2\psi}{d\sigma^2} + \frac{2}{r}\frac{dr}{d\sigma}\frac{d\psi}{d\sigma} + 2\cot\theta\frac{d\psi}{d\sigma}\frac{d\theta}{d\sigma} = 0,$$

$$\frac{d^2 t}{d\sigma^2} + \frac{2m}{c^2 r^2}\left(1 - \frac{2m}{c^2 r}\right)^{-1}\frac{dr}{d\sigma}\frac{dt}{d\sigma} = 0,$$

$$-\left(1 - \frac{2m}{c^2 r}\right)^{-1}\left(\frac{dr}{d\sigma}\right)^2 - r^2\left(\frac{d\theta}{d\sigma}\right)^2 - r^2\sin^2\theta\left(\frac{d\psi}{d\sigma}\right)^2 + c^2\left(1 - \frac{2m}{c^2 r}\right)\left(\frac{dt}{d\sigma}\right)^2 = 1.$$

We may assume that the planet moves initially in the plane $\theta = \pi/2$. That is, $d\theta/d\sigma$ and $\cos\theta$ are both initially zero. Then (69.1) tells us that $d^2\theta/d\sigma^2$ is also zero. Repeated differentiation of this equation shows that $d^i\theta/d\sigma^i$ vanishes at $t = 0$ for all i. Hence $\theta = \pi/2$ permanently, and the above equations simplify to

$$\frac{d^2\psi}{d\sigma^2} + \frac{2}{r}\frac{dr}{d\sigma}\frac{d\psi}{d\sigma} = 0, \tag{69.2}$$

$$\frac{d^2t}{d\sigma^2} + \frac{2m}{c^2r}\left(1 - \frac{2m}{c^2r}\right)^{-1}\frac{dr}{d\sigma}\frac{dt}{d\sigma} = 0, \tag{69.3}$$

$$-\left(1-\frac{2m}{c^2r}\right)^{-1}\left(\frac{dr}{d\sigma}\right)^2 - r^2\left(\frac{d\psi}{d\sigma}\right)^2 + c^2\left(1-\frac{2m}{c^2r}\right)\left(\frac{dt}{d\sigma}\right)^2 = 1. \tag{69.4}$$

We can immediately integrate (69.2) and (69.3) and the results are

$$r^2\frac{d\psi}{d\sigma} = h, \quad \left(1 - \frac{2m}{c^2r}\right)\frac{dt}{d\sigma} = k, \tag{69.5}$$

where h and k are constants. On eliminating t and σ from (69.4) and (69.5) we obtain

$$-\frac{1}{r^4}\left(\frac{dr}{d\psi}\right)^2 - \frac{1}{r^2}\left(1 - \frac{2m}{c^2r}\right) + \frac{c^2k^2}{h^2} = \frac{1}{h^2}\left(1 - \frac{2m}{c^2r}\right).$$

Now substitute $r = 1/u$ and differentiate the equation so obtained with respect to ψ and the result is

$$\frac{d^2u}{d\psi^2} + u = \frac{m}{c^2h^2} + \frac{3mu^2}{c^2}. \tag{69.6}$$

For the planets of our solar system, the term m/c^2h^2 is much larger than $3mu^2/c^2$. But when we neglect this latter term, we obtain Newton's equation for the motion of a planet. Thus the first approximation to the solution of (69.6) is the Newtonian solution $u = \dfrac{m}{c^2h^2}\{1+e\cos(\psi-\xi)\}$, where e is the eccentricity of the elliptic orbit and ξ is the longitude of perihelion. A second approximation to the solution can then be obtained in the form $u = \frac{m}{c^2h^2}\{1 + e\cos(\psi - \xi - \Delta\xi)\}$, where $\Delta\xi = 3m^2\psi/c^4h^2$. This means that the major axis of the elliptic orbit is slowly rotating about its focus (the sun). The increase in $\Delta\xi$ corresponding to a complete revolution $\psi = 2\pi$ is thus $6m^2\pi/c^4h^2$. For the planet Mercury the advance of perihelion is calculated from this to be 42.9 seconds of arc per century. This agrees well with the observational figures of 43.5 seconds of arc per century.

§ 70. Einstein's universe

Einstein was led by cosmological considerations to consider the universe with the metric

$$(70.1)\quad d\sigma^2 = -(1 - r^2/\mathscr{R}^2)^{-1}dr^2 - r^2d\theta^2 - r^2\sin^2\theta d\psi^2 + c^2dt^2,$$

where $\mathscr{R}$ is a constant. This metric is spherically symmetrical with $e^{-\lambda} = (1 - r^2/\mathscr{R}^2)$ and $\nu = 0$. The Christoffel symbols of the second kind are readily obtained from (67.2).

Let us investigate the path of a ray of light in Einstein's universe. The path must be a null-geodesic and so its equations are given by three of the four equations (26.4) taken together with $g_{ij}\frac{dx^i}{du}\frac{dx^j}{du} = 0$. That is, we have

$$(70.2)\quad \frac{d^2\theta}{du^2} + \frac{2}{r}\frac{d\theta}{du}\frac{dr}{du} - \sin\theta\cos\theta\left(\frac{d\psi}{du}\right)^2 = 0,$$

$$\frac{d^2\psi}{du^2} + \frac{2}{r}\frac{d\psi}{du}\frac{dr}{du} + 2\cot\theta\frac{d\psi}{du}\frac{d\theta}{du} = 0,$$

$$\frac{d^2t}{du^2} = 0,$$

$$-(1 - r^2/\mathscr{R}^2)^{-1}\left(\frac{dr}{du}\right)^2 - r^2\left(\frac{d\theta}{du}\right)^2 - r^2\sin^2\theta\left(\frac{d\psi}{du}\right)^2 + c^2\left(\frac{dt}{du}\right)^2 = 0,$$

where u is some parameter. Again, following the argument of section 69, the equation (70.2) tells us that we can take θ to have the permanent value $\pi/2$. With this choice, the remaining equations reduce to

$$\frac{d^2\psi}{du^2} + \frac{2}{r}\frac{d\psi}{du}\frac{dr}{du} = 0, \qquad \frac{d^2t}{du^2} = 0, \tag{70.3}$$

$$-(1-r^2/\mathscr{R}^2)^{-1}\left(\frac{dr}{du}\right)^2 - r^2\left(\frac{d\psi}{du}\right)^2 + c^2\left(\frac{dt}{du}\right)^2 = 0. \tag{70.4}$$

We integrate (70.3) and obtain

$$r^2\frac{d\psi}{du} = h, \qquad \frac{dt}{du} = k, \tag{70.5}$$

where h and k are constants, and then eliminate t and u from these equations and (70.4). The result is

$$\left(\frac{dr}{d\psi}\right)^2 = r^2\left(1 - \frac{r^2}{\mathscr{R}^2}\right)\left(\frac{c^2k^2}{h^2}r^2 - 1\right).$$

The solution of this equation is

$$\frac{1}{r^2} = \frac{1}{\mathscr{R}^2}\cos^2(\psi - \xi) + \frac{c^2k^2}{h^2}\sin^2(\psi - \xi) \tag{70.6}$$

where ξ is a constant. We immediately see that r regains its initial value when ψ is increased by π and that r is never infinite for any value of ψ. Thus all the null-geodesics of Einstein's universe, that is the light rays, are closed curves. We see from (70.5) that $\dfrac{dt}{d\psi} = \dfrac{k}{h}r^2$. Hence the time taken for a light ray to make a complete circuit is given by

$$T = \frac{k}{h}\int_0^{2\pi}\left\{\frac{1}{\mathscr{R}^2}\cos^2(\psi - \xi) + \frac{c^2k^2}{h^2}\sin^2(\psi - \xi)\right\}^{-1} d\psi.$$

Owing to the periodicity of ψ we have

$$T = \frac{k}{h}\int_0^{2\pi}\left\{\frac{1}{\mathscr{R}^2}\cos^2\psi + \frac{c^2k^2}{h^2}\sin^2\psi\right\}^{-1} d\psi$$

$$= \frac{4k}{h}\int_0^{\pi/2}\left\{\frac{1}{\mathscr{R}^2}\cos^2\psi + \frac{c^2k^2}{h^2}\sin^2\psi\right\}^{-1} d\psi$$

and on carrying out the integration we find that $T = 2\pi\mathscr{R}/c$.

Ex. 1. Show that the Einstein universe is neither an Einstein space nor a space of constant curvature.

Ex. 2. Prove that the curvature invariant of Einstein's universe is $R = 6/\mathscr{R}^2$.

§ 71. De Sitter's universe

Other cosmological considerations suggested to De Sitter that the universe could be described by the metric

$$d\sigma^2 = -(1 - r^2/\mathscr{R}^2)^{-1}\,dr^2 - r^2\,d\theta^2 - r^2\sin^2\theta\,d\psi^2 + c^2(1 - r^2/\mathscr{R}^2)dt^2. \tag{71.1}$$

This metric has also spherical symmetry but the constant $\mathscr{R}$ has not the same value as the corresponding constant of the Einstein universe.

The paths of light rays are the null-geodesics given by the equations

$$\frac{d^2\theta}{du^2} + \frac{2}{r}\frac{dr}{du}\frac{d\theta}{du} - \sin\theta\cos\theta\left(\frac{d\psi}{du}\right)^2 = 0,$$

$$\frac{d^2\psi}{du^2} + \frac{2}{r}\frac{dr}{du}\frac{d\psi}{du} + 2\cot\theta\,\frac{d\theta}{du}\frac{d\psi}{du} = 0, \tag{71.2}$$

$$\frac{d^2t}{du^2} - \frac{2r}{\mathscr{R}^2}(1 - r^2/\mathscr{R}^2)^{-1}\frac{dt}{du}\frac{dr}{du} = 0, \tag{71.3}$$

$$-(1 - r^2/\mathscr{R}^2)^{-1}\left(\frac{dr}{du}\right)^2 - r^2\left(\frac{d\theta}{du}\right)^2 - r^2\sin^2\theta\left(\frac{d\psi}{du}\right)^2 + c^2(1 - r^2/\mathscr{R}^2)\left(\frac{dt}{du}\right)^2 = 0. \tag{71.4}$$

Again we can choose $\theta = \pi/2$ permanently, and on integration the equations (71.2) and (71.3) become

$$r^2 \frac{d\psi}{du} = h, \quad \frac{dt}{du} = k(1 - r^2/\mathscr{R}^2)^{-1}.$$

We now eliminate t and u from these equations and (71.4). The result is

$$\left(\frac{dr}{d\psi}\right)^2 = r^2(a^2 r^2 - 1)$$

where $a^2 = c^2k^2/h^2 + 1/\mathscr{R}^2$. This equation can be immediately integrated and yields

$$\frac{1}{r} = a \cos (\psi - \xi), \tag{71.5}$$

where ξ is a constant. These trajectories correspond to straight lines and are not closed, since r becomes infinite when $\psi - \xi = \pi/2$.

Ex. Show that the De Sitter universe is an Einstein space with constant curvature $12/\mathscr{R}^2$.

Solution

p. 114. Ex. $e^{-\lambda} = 1 + ar^2 + b/r$; $\nu = k - \lambda$, where a, b and k are constants.

BIBLIOGRAPHY

TENSOR THEORY

J. A. Schouten, Der Ricci-Kalkül, (1924).

L. P. Eisenhart, Riemannian Geometry, (1926).

* T. Levi-Civita, The Absolute Differential Calculus, (1927).

O. Veblen, Invariants of Quadratic Differential Forms, (1927).

H. Jeffreys, Cartesian Tensors, (1931).

A. J. McConnell, Applications of the Absolute Differential Calculus, (1931).

J. A. Schouten and D. J. Struik, Einführung in die Neueren Methoden der Differentialgeometrie, Vol. 1 (1935), Vol. 2 (1938).

C. E. Weatherburn, Riemannian Geometry and the Tensor Calculus, (1938).

A. D. Michal, Matrix and Tensor Calculus, (1947).

L. Brand, Vector and Tensor Calculus, (1947).

* J. L. Synge and A. Schild, Tensor Calculus, (1949).

* J. A. Schouten, Tensor Calculus for Physicists, (1951).

J. A. Schouten, Ricci-Calculus (1954).

A detailed list of original memoirs on Tensor Theory is contained in Vol. 2 of J. A. Schouten and D. J. Struik's book and in Schouten (1954).

DIFFERENTIAL GEOMETRY

L. P. Eisenhart, Treatise on the Differential Geometry of Curves and Surfaces, (1909).

W. Blaschke, Vorlesungen über Differentialgeometrie, Vol. 1, 2nd edition, (1924), Vol. 2 (1923), Vol. 3 (1929).

C. E. Weatherburn, Differential Geometry of 3 Dimensions, Vol. 1 (1927), Vol. 2 (1930).

ELASTICITY

* A. E. H. Love, Treatise on the Mathematical Theory of Elasticity, 4th. edition, (1927).

I. S. Sokolnikoff, Mathematical Theory of Elasticity, (1946).

L. Brillouin, Les tenseurs en mécanique et en elasticite, (1946).

RELATIVITY

* H. Weyl, Space-Time-Matter, (1922).

* A. Einstein, The Principle of Relativity, (1923).

A. S. Eddington, Mathematical Theory of Relativity, (1923).

* R. C. Tolman, Relativity, Thermodynamics and Cosmology, (1934).

* P. G. Bergmann, Introduction to the Theory of Relativity, (1942).

A. Einstein, The Meaning of Relativity, 4th edition, (1950).

* W. Pauli, Theory of Relativity, (1958).

W. Rindler, Special Relativity, (in preparation).

* Republished by Dover Publications, Inc. For availability and current prices log on to www.doverpublications.com.

INDEX

The numbers refer to pages.

A CATALOG OF SELECTED

DOVER BOOKS

IN SCIENCE AND MATHEMATICS

Astronomy

BURNHAM'S CELESTIAL HANDBOOK, Robert Burnham, Jr. Thorough guide to the stars beyond our solar system. Exhaustive treatment. Alphabetical by constellation: Andromeda to Cetus in Vol. 1; Chamaeleon to Orion in Vol. 2; and Pavo to Vulpecula in Vol. 3. Hundreds of illustrations. Index in Vol. 3. 2,000pp. 6⅛ x 9¼.
23567-X, 23568-8, 23673-0 Three-vol. set

THE EXTRATERRESTRIAL LIFE DEBATE, 1750–1900, Michael J. Crowe. First detailed, scholarly study in English of the many ideas that developed from 1750 to 1900 regarding the existence of intelligent extraterrestrial life. Examines ideas of Kant, Herschel, Voltaire, Percival Lowell, many other scientists and thinkers. 16 illustrations. 704pp. 5⅜ x 8½. 40675-X

A HISTORY OF ASTRONOMY, A. Pannekoek. Well-balanced, carefully reasoned study covers such topics as Ptolemaic theory, work of Copernicus, Kepler, Newton, Eddington's work on stars, much more. Illustrated. References. 521pp. 5⅜ x 8½.
65994-1

AMATEUR ASTRONOMER'S HANDBOOK, J. B. Sidgwick. Timeless, comprehensive coverage of telescopes, mirrors, lenses, mountings, telescope drives, micrometers, spectroscopes, more. 189 illustrations. 576pp. 5⅝ x 8¼. (Available in U.S. only.)
24034-7

STARS AND RELATIVITY, Ya. B. Zel'dovich and I. D. Novikov. Vol. 1 of *Relativistic Astrophysics* by famed Russian scientists. General relativity, properties of matter under astrophysical conditions, stars, and stellar systems. Deep physical insights, clear presentation. 1971 edition. References. 544pp. 5⅝ x 8¼. 69424-0

Chemistry

CHEMICAL MAGIC, Leonard A. Ford. Second Edition, Revised by E. Winston Grundmeier. Over 100 unusual stunts demonstrating cold fire, dust explosions, much more. Text explains scientific principles and stresses safety precautions. 128pp. 5⅜ x 8½. 67628-5

THE DEVELOPMENT OF MODERN CHEMISTRY, Aaron J. Ihde. Authoritative history of chemistry from ancient Greek theory to 20th-century innovation. Covers major chemists and their discoveries. 209 illustrations. 14 tables. Bibliographies. Indices. Appendices. 851pp. 5⅜ x 8½. 64235-6

CATALYSIS IN CHEMISTRY AND ENZYMOLOGY, William P. Jencks. Exceptionally clear coverage of mechanisms for catalysis, forces in aqueous solution, carbonyl- and acyl-group reactions, practical kinetics, more. 864pp. 5⅜ x 8½.
65460-5

Math–Geometry and Topology

ELEMENTARY CONCEPTS OF TOPOLOGY, Paul Alexandroff. Elegant, intuitive approach to topology from set-theoretic topology to Betti groups; how concepts of topology are useful in math and physics. 25 figures. 57pp. 5⅜ x 8½. 60747-X

COMBINATORIAL TOPOLOGY, P. S. Alexandrov. Clearly written, well-organized, three-part text begins by dealing with certain classic problems without using the formal techniques of homology theory and advances to the central concept, the Betti groups. Numerous detailed examples. 654pp. 5⅜ x 8½. 40179-0

EXPERIMENTS IN TOPOLOGY, Stephen Barr. Classic, lively explanation of one of the byways of mathematics. Klein bottles, Moebius strips, projective planes, map coloring, problem of the Koenigsberg bridges, much more, described with clarity and wit. 43 figures. 210pp. 5⅜ x 8½. 25933-1

CONFORMAL MAPPING ON RIEMANN SURFACES, Harvey Cohn. Lucid, insightful book presents ideal coverage of subject. 334 exercises make book perfect for self-study. 55 figures. 352pp. 5⅝ x 8¼. 64025-6

THE GEOMETRY OF RENÉ DESCARTES, René Descartes. The great work founded analytical geometry. Original French text, Descartes's own diagrams, together with definitive Smith-Latham translation. 244pp. 5⅜ x 8½. 60068-8

THE THIRTEEN BOOKS OF EUCLID'S ELEMENTS, translated with introduction and commentary by Sir Thomas L. Heath. Definitive edition. Textual and linguistic notes, mathematical analysis. 2,500 years of critical commentary. Unabridged. 1,414pp. 5⅜ x 8½. Three-vol. set.
Vol. I: 60088-2 Vol. II: 60089-0 Vol. III: 60090-4

GEOMETRY OF COMPLEX NUMBERS, Hans Schwerdtfeger. Illuminating, widely praised book on analytic geometry of circles, the Moebius transformation, and two-dimensional non-Euclidean geometries. 200pp. 5⅝ x 8¼. 63830-8

DIFFERENTIAL GEOMETRY, Heinrich W. Guggenheimer. Local differential geometry as an application of advanced calculus and linear algebra. Curvature, transformation groups, surfaces, more. Exercises. 62 figures. 378pp. 5⅜ x 8½. 63433-7

CURVATURE AND HOMOLOGY: Enlarged Edition, Samuel I. Goldberg. Revised edition examines topology of differentiable manifolds; curvature, homology of Riemannian manifolds; compact Lie groups; complex manifolds; curvature, homology of Kaehler manifolds. New Preface. Four new appendixes. 416pp. 5⅜ x 8½. 40207-X

TOPOLOGY, John G. Hocking and Gail S. Young. Superb one-year course in classical topology. Topological spaces and functions, point-set topology, much more. Examples and problems. Bibliography. Index. 384pp. 5⅝ x 8¼. 65676-4

Physics

OPTICAL RESONANCE AND TWO-LEVEL ATOMS, L. Allen and J. H. Eberly. Clear, comprehensive introduction to basic principles behind all quantum optical resonance phenomena. 53 illustrations. Preface. Index. 256pp. 5⅜ x 8½. 65533-4

ULTRASONIC ABSORPTION: An Introduction to the Theory of Sound Absorption and Dispersion in Gases, Liquids and Solids, A. B. Bhatia. Standard reference in the field provides a clear, systematically organized introductory review of fundamental concepts for advanced graduate students, research workers. Numerous diagrams. Bibliography. 440pp. 5⅜ x 8½. 64917-2

QUANTUM THEORY, David Bohm. This advanced undergraduate-level text presents the quantum theory in terms of qualitative and imaginative concepts, followed by specific applications worked out in mathematical detail. Preface. Index. 655pp. 5⅜ x 8½. 65969-0

ATOMIC PHYSICS (8th edition), Max Born. Nobel laureate's lucid treatment of kinetic theory of gases, elementary particles, nuclear atom, wave-corpuscles, atomic structure and spectral lines, much more. Over 40 appendices, bibliography. 495pp. 5⅜ x 8½. 65984-4

AN INTRODUCTION TO HAMILTONIAN OPTICS, H. A. Buchdahl. Detailed account of the Hamiltonian treatment of aberration theory in geometrical optics. Many classes of optical systems defined in terms of the symmetries they possess. Problems with detailed solutions. 1970 edition. xv + 360pp. 5⅜ x 8½. 67597-1

THIRTY YEARS THAT SHOOK PHYSICS: The Story of Quantum Theory, George Gamow. Lucid, accessible introduction to influential theory of energy and matter. Careful explanations of Dirac's anti-particles, Bohr's model of the atom, much more. 12 plates. Numerous drawings. 240pp. 5⅜ x 8½. 24895-X

ELECTRONIC STRUCTURE AND THE PROPERTIES OF SOLIDS: The Physics of the Chemical Bond, Walter A. Harrison. Innovative text offers basic understanding of the electronic structure of covalent and ionic solids, simple metals, transition metals and their compounds. Problems. 1980 edition. 582pp. 6⅛ x 9¼. 66021-4

HYDRODYNAMIC AND HYDROMAGNETIC STABILITY, S. Chandrasekhar. Lucid examination of the Rayleigh-Benard problem; clear coverage of the theory of instabilities causing convection. 704pp. 5⅜ x 8¼. 64071-X

INVESTIGATIONS ON THE THEORY OF THE BROWNIAN MOVEMENT, Albert Einstein. Five papers (1905–8) investigating dynamics of Brownian motion and evolving elementary theory. Notes by R. Fürth. 122pp. 5⅜ x 8½. 60304-0

THE PHYSICS OF WAVES, William C. Elmore and Mark A. Heald. Unique overview of classical wave theory. Acoustics, optics, electromagnetic radiation, more. Ideal as classroom text or for self-study. Problems. 477pp. 5⅜ x 8½. 64926-1

CATALOG OF DOVER BOOKS

METHODS OF THERMODYNAMICS, Howard Reiss. Outstanding text focuses on physical technique of thermodynamics, typical problem areas of understanding, and significance and use of thermodynamic potential. 1965 edition. 238pp. 5⅜ x 8½. 69445-3

TENSOR ANALYSIS FOR PHYSICISTS, J. A. Schouten. Concise exposition of the mathematical basis of tensor analysis, integrated with well-chosen physical examples of the theory. Exercises. Index. Bibliography. 289pp. 5⅜ x 8½. 65582-2

RELATIVITY IN ILLUSTRATIONS, Jacob T. Schwartz. Clear nontechnical treatment makes relativity more accessible than ever before. Over 60 drawings illustrate concepts more clearly than text alone. Only high school geometry needed. Bibliography. 128pp. 6⅛ x 9¼. 25965-X

THE ELECTROMAGNETIC FIELD, Albert Shadowitz. Comprehensive undergraduate text covers basics of electric and magnetic fields, builds up to electromagnetic theory. Also related topics, including relativity. Over 900 problems. 768pp. 5⅜ x 8¼. 65660-8

GREAT EXPERIMENTS IN PHYSICS: Firsthand Accounts from Galileo to Einstein, edited by Morris H. Shamos. 25 crucial discoveries: Newton's laws of motion, Chadwick's study of the neutron, Hertz on electromagnetic waves, more. Original accounts clearly annotated. 370pp. 5⅜ x 8½. 25346-5

RELATIVITY, THERMODYNAMICS AND COSMOLOGY, Richard C. Tolman. Landmark study extends thermodynamics to special, general relativity; also applications of relativistic mechanics, thermodynamics to cosmological models. 501pp. 5⅜ x 8½. 65383-8

LIGHT SCATTERING BY SMALL PARTICLES, H. C. van de Hulst. Comprehensive treatment including full range of useful approximation methods for researchers in chemistry, meteorology and astronomy. 44 illustrations. 470pp. 5⅜ x 8½. 64228-3

STATISTICAL PHYSICS, Gregory H. Wannier. Classic text combines thermodynamics, statistical mechanics and kinetic theory in one unified presentation of thermal physics. Problems with solutions. Bibliography. 532pp. 5⅜ x 8½. 65401-X